P9-CLB-096

THE JOKE'S OVER

ALSO BY

RALPH STEADMAN

Gonzo

The Book of Jones

Still Life with Bottle

The Grapes of Ralph

Untrodden Grapes

BRUISED MEMORIES: GONZO, HUNTER S. THOMPSON, AND ME

A Harvest Book
Harcourt, Inc.
Orlando Austin New York San Diego London

Copyright © Ralph Steadman, 2006
Foreword copyright © Kurt Vonnegut, 2006

All rights reserved. No part of this publication may be reproduced
or transmitted in any form or by any means, electronic or mechanical,
including photocopy, recording, or any information storage and retrieval
system, without permission in writing from the publisher.

Requests for permission to make copies of any part of the work should be
submitted online at www.harcourt.com/contact or mailed to the following
address: Permissions Department, Harcourt, Inc., 6277 Sea Harbor Drive,
Orlando, Florida 32887-6777.

www.HarcourtBooks.com

Letters and faxes written by Hunter S. Thompson and addressed to
Ralph Steadman are reproduced by kind permission of Doug Brinkley
on behalf of the Estate of Hunter S. Thompson.

Published in Great Britain by William Heinemann

The Library of Congress has cataloged the hardcover edition as follows:
Steadman, Ralph.
The joke's over: bruised memories: Gonzo, Hunter Thompson and me /
Ralph Steadman.—1st U.S. ed.
p. cm.
1. Steadman, Ralph. 2. Cartoonists—Great Britain—Biography 3. Steadman,
Ralph—Friends and associates. 4. Thompson, Hunter S. I. Title.
NC1479.S79A2   2006
741.5092—dc22 [B]   2006011375
ISBN 978-0-15-101282-4
ISBN 978-0-15-603250-6 (pbk.)

Text set in Walbaum MT

Printed in the United States of America

First Harvest edition 2007
A C E G I K J H F D B

For ANNA, centre of my universe
and Nat Sobel in orbit
and it's for you too, you ole BASTARD!
wherever you are!

CONTENTS

LIST OF PLATES
x

ACKNOWLEDGEMENTS
xv

FOREWORD
xvii

INTRODUCTION
1

THE SEVENTIES
5

THE KENTUCKY DERBY. MAY 1970
6

THE AMERICA'S CUP. SEPTEMBER 1970
35

FEAR AND LOATHING. SUMMER 1971
66

HUNTER GOES TO WASHINGTON.
1970–72
78

WATERGATE FOLLIES. JULY 1973
92

RUMBLE IN THE JUNGLE: ALI v
FOREMAN. 30 OCTOBER 1974
115

FEAR AND LOATHING ON THE ROAD
TO HOLLYWOOD. 1977
137

THE FILMING OF *WHERE THE BUFFALO
ROAM*. 1979
167

THE EIGHTIES
183

'THE EIGHTIES, RALPH, ARE ABOUT
PAYING YOUR RENT.' 1980
184

HAWAII. 1980
195

THE ELUSIVE BLOATER. 1981
206

TAXI TO THE SUBURBS. 1981
237

THE FISH HAVE GONE SOUTH. 1981–6
245

THE YEAR OF WINE. 1987
271

THE NINETIES
275

OWN GOALS. 1990–94
276

WILLIAM BURROUGHS – AN
ENCOUNTER. 1995
325

RIFLE. MARCH 1996
338

VISIT TO VIRGINIA, HUNTER'S
MOTHER. LOUISVILLE, 1997
344

THE END
349

THE RED SHARK. 2000
350

READING *LONO* OUT LOUD. 2000
359

THE LAST TRIP TO WOODY CREEK.
2004
366

MEMO TO THE SPORTS DESK. 2006
383

INDEX
388

LIST OF PLATES

FIRST SECTION
Page 1: 1970. Scenes of the Kentucky Derby, including showing Ralph in toilet mirror, Warren Hinckle, with eye patch, and Sidney Zion (editors of *Scanlan's*), the press room at the Derby and spectators.
1970. Drawing of search for the artist to be used to illustrate *Fear and Loathing in Las Vegas*.

Page 2: Original spread for first instalment of *Fear and Loathing in Las Vegas* which appeared in *Rolling Stone* magazine in November 1971.
Scene from *Fear and Loathing in Las Vegas* of the police chiefs at a drugs convention.
1977. Hunter and Ralph in the desert between scenes for the film *Fear and Loathing on the Road to Hollywood*.

Page 3: Dr Jo Sterling Baxter, Republican candidate for Pitkin County Commissioner and Hunter opponent during his sheriff 'freak power' campaign.
Ralph's drawing for Hunter's campaign for sheriff.
The rest of Hunter's opposition: A. Incumbent sheriff Carrol Whitemore, his Democratic opponent. B. Glen Ricks, Republican candidate opponent.

Page 4: 1972. Scenes of fishing picnic at the Roaring Fork River, Woody Creek, Aspen, Colorado, with Ralph, Anna and Juan Thompson, Hunter's son.

Page 5: 1972. Intimate studies of Hunter and Sandy at Owl Farm, Woody Creek.
Patriotic cabin on road to Woody Creek.
Front entrance of Owl Farm.
2005. Memorial flag at Owl Farm at sunset.

Page 6: 1973. Hunter poolside at the Hilton Hotel, Washington D.C., during the Watergate hearings
1981. Hunter at the Sheraton Hotel, Washington D.C.

Page 7: 1973. Hunter's first meeting with Mo Dean, Beverly Hills.
1973. Art Garfunkel and Jann Wenner, San Francisco.
1976. Bill Cardoso, the man who coined the word 'gonzo', in Sausalito, California.

Page 8: 1973. Anna in the regalia of a Navaho squaw during the Santa Fe Indian Festival, and Ralph in a Navaho hat.
1976. Jann Wenner and John Dean in the hall at the Republican Convention, Kansas City.
John Dean and Ralph in Washington D.C.

SECOND SECTION
Page 1: 1974. Unpublished drawing of Hunter and Ralph as voodoo dolls intended for the never-published *Rolling Stone* issue on the Ali–Foreman fight, described as 'the biggest fucked-up story in the history of journalism'.

Page 2: 1974. Intercontinental Hotel, Kinshasa, Zaire. Hunter relaxes on hotel bed before the Ali-Foreman fight, known as 'The Rumble in the Jungle'.
Hunter in hotel pool.
Ralph in hotel room.
Hunter negotiates purchase of elephant tusks with $300 of non-negotiable American Express travellers' cheques.

Page 3: 1974. George Foreman press conference, Zaire.
Hunter's study of Ralph levitating in pool at Intercontinental Hotel, Zaire.
Voodoo doll, Zaire.
Hunter phoning America from Zaire.

Page 4: 1979. L'Hermitage Hotel, LA, during a rewrite of the introduction to the film *Where the Buffalo Roam*.
Ralph fools around in front of his graphics.
Hunter manipulates electronic machinery in hotel room.

Page 5: 1979. Studies of Bill Murray during the filming of *Where the Buffalo Roam*.

Page 6: 1981. The fishing tournament, Kona. Ralph in the fighting chair on the *Haere Maru*.

Ralph and Wild Turkey.

Hunter with son Juan on board the *Haere Maru*.

Captain Steve, owner of the *Haere Maru*.

Page 7: 1981. Hunter on board the *Haere Maru*.

Hunter talks to Captain Steve.

Hunter, deep in thought, considers our dilemma.

Hunter reads the paper during the lull.

All at sea being towed in by the *Little Child*.

Page 8: 1981. Drawing area set up at Owl Farm to complete the drawings for *The Curse of Lono*.

1980. Hunter watching runners at the Honolulu Marathon.

'Nuclear War — Race Cancelled' — Ralph's chalked message to marathon runners at party held at the bottom of Heartbreak Hill.

THIRD SECTION

Page 1: 1980. Anna and daughter Sadie in the pool on the Kona Coast, Hawaii.

Anna and Sadie pose on the diving board of the pool overlooking the Pacific.

Dramatic Kona coastline.

Laila relaxes by the pool at the compound.

Sadie dressed for the Christmas party, December 1980.

Ralph serenades on the verandah of the compound.

Page 2: 1980. Hunter reflects at the edge of the City of Refuge on the island of Kona, Hawaii.

Totem effigies protecting the City of Refuge from the outside world.

Ralph as 'buggering fool' outside the City of Refuge.

Ralph taking pictures around City of Refuge.

Makeshift drawing board and anti-splash screen at the Kona compound where drawings were conceived for the book *The Curse of Lono*.

Page 3: 1981. Hunter catches his fish on his second trip to Kona.

Hunter sets the rod.

The catch on board.

Page 4: 1994. Anna makes friends with Hunter's stuffed wolverine in the back of the Red Shark.

Nocturnal visit in Aspen rental house. Hunter attacks Anna with his toy hammer.

Ralph and Hunter in limousine on the way to Juan's wedding, in the hills outside Boulder, Colorado.
Hunter adjusts his son's bow tie prior to the wedding.

Page 5: 1994. Lawrence, Kansas. Ralph showing William Burroughs his art. Photo credit Joe Petro III.
All pals together: William, Anna and Ralph. Photo credit Joe Petro III.
Ralph shooting and William shooting. Photo credit Joe Petro III.

Page 6: 1994. Gun-blasted ink on paper used to make stencil for Sheriff print laid out on the trunk of the Red Shark.
1995. Joe Petro watches Hunter sign the limited edition of the Sheriff print.
Sheriff shotgun art. Detail of Sheriff print showing blast.

Page 7: Hunter's weapons arsenal, War Room, Owl Farm.
1996. Hunter places the blow-up doll on his tractor prior to shooting a propane gas cylinder 'bomb'. Photo credit Joe Petro III.
The bomb explodes.
Ralph and Hunter getting ready to shoot ink bottles at artwork. Photo credit Joe Petro III.

Page 8: 1994. Hunter visits the Steadmans at their rented house in Aspen and offers counsel.
1996. Owl Farm. Hunter gives Ralph, dressed as a woman, a rare embrace.
Ralph in performance – drag act.
Hunter signs Flying Dog Ale poster.
Hunter uses electric-shock gadget to warn off excessive video-filming.

FOURTH SECTION
Page 1: 1996. The Lotus Club, New York. The *Rolling Stone* twenty-fifth anniversary party to celebrate *Fear and Loathing in Las Vegas*, first published in the magazine in 1971; Laila Nabulsi, Johnny Depp, Ralph and Kate Moss.
Hunter and Aspen friend, Cilla Hyams.
The Four Seasons Hotel, New York. Laila comforts a dead-beat Hunter after the celebration.
Ralph, Jann Wenner, Janie Wenner and Hunter.

Page 2: 1995. Hunter and Ralph cutting the cake to celebrate the twenty-fifth anniversary of the birth of Gonzo in the kitchen at Owl Farm, Woody Creek.

Studies of Hunter in his fighting chair at his kitchen counter, Owl Farm.

Page 3: 2004. Scenes in the kitchen at Owl Farm and discussions about the unfinished *Polo is My Life* with Belinda, the four-eyed horse over the piano.

Page 4: Virginia Thompson, Hunter's mother, aged ninety-two, in her room at her care home, Louisville, Kentucky, surrounded by her family photos and paintings during Ralph's visit in 1997.

Page 5: 2004. Ralph working on the title *Vote Naked and Die* at George Stranahan's ranch house.
Oxygen for Ralph.
A spliff for the pig at the Woody Creek Tavern.
Kangaroo court poster at Owl Farm.
Hunter with good friend Sheriff Bob Braudis.
The cheque that Ralph gave to Hunter for signing the limited edition of his last book, *Fire in the Nuts*.

Page 6: 2005. The Jerome Hotel, Aspen. Johnny Depp signs fist monument print based on the original drawing created by Ralph in 1977 and printed by Joe Petro, and to be sold in aid of the Hunter S. Thompson estate.
Bill Murray shows photos of his early acting career.
Ralph impersonating Hunter.

Page 7: 2005. Ralph with Anita Thompson, Hunter's wife, just after the March memorial.
Ralph with Laila Nabulsi in Owl Farm kitchen.
1979. Drawing of Hunter from the back jacket of *The Great Shark Hunt*.

Page 8: 2005. The fist monument, Woody Creek. First conceived in 1977 in a West Hollywood funeral parlour which Hunter asked Ralph to design. Joe Petro, Anna and Ralph Steadman seated beneath the realized monument, funded by Johnny Depp and activated to send Hunter's ashes into the firmament on 20 August 2005.

ACKNOWLEDGEMENTS

Special thanks to Robert Chalmers for invaluable access to his London notes on the visit to London of Hunter S. Thompson in 1992, for the *Observer*, knowing as I do how much pain went into collecting such information, physical risk, endless personal contempt and derision, and some virulent fear and loathing from his prey. 'The only journalist I have ever resonated with', said Hunter after a Chalmers interview.

It has been my good fortune to know a great American writer and one of his country's strongest and perceptive critics, Kurt Vonnegut. I value the generosity of the words he has contributed to my book in his foreword. I am honored that he is my friend – and funny too!

To Caroline Knight for her tireless enthusiasm in bringing together the wandering bits of my mind and sternly warning them into place within the parameters of what most sensible people refer to as a book, like a St Trinian's headmistress on a night out and smoking my fags.

To Simon McFadden for the elegance of his typography and the quiet acceptance of a volatile, writhing object, identifying it as a manuscript, wrestling it to the ground and bringing it to heel.

To Mark Handsley, for the perspicacity of his copy-editing and the eye-boggling shuffle of typewritten sheets of paper, reminiscent of a card croupier in the Mandalay Bay Casino in Las Vegas.

To Ravi Mirchandani, my publisher, for his utter confidence in my ability to pull it off, gather up the bits to his bosom and accept it for legitimate adoption under his own roof.

To Andre Bernard, my American publisher, who has simultaneously adopted this mutant bastard and given it US citizenship, enabling it to roam free and uncensored, anywhere it damn well likes by the sound of things!

To Doug Brinkley, Juan Thompson, son of, and Anita Thompson, wife of, and Trustees of the Hunter S. Thompson Foundation for allowing me to use parts of my correspondence between me and my late friend, enabling me to build up a strong case against him for thirty-five years of verbal abuse and criminal usury.

To Cassie Chadderton, brutal and ruthless administrator of my book tour, whose earthly task has been to whip up a frenzy of excitement about just what the world needs – another book!

To Henry Steadman, my son, who designed the covers for both the UK and American editions, offering a superb array of possibilities and changes in execution during a period when deciding on a title resembled a Bookmaker's Betting Board five minutes before the big race more than a serious in-house editorial meeting.

And to my editor, Gordon Kerr, 'The World's Greatest Living Poet'. Every time, he makes stuff rhyme. Check him out!

FOREWORD

by Kurt Vonnegut

This good book is by a friend of mine, a Welshman who, in my humble opinion, has been the most gifted and effective existentialist graphic artist of my time. It includes within a modest and appealing memoir a portrait in well-chosen words of an American journalist, a native of Kentucky and also a friend of mine, whose exceedingly personal, desperately brilliant writings Ralph Steadman illustrated for more than a third of a century. I mean the gun nut and drug abuser and heavy consumer of grain alcohol, and finally a gun suicide, Hunter S. Thompson (1937–2005).

Until I myself read and then met Hunter, I would have thought it impossible for anyone whose brains were so saturated with mind-benders to make sense on a telephone, let alone write so well. In revolt against life itself all his life, may Hunter rest in peace at last, now that his ashes have, at his own request, been shot into the sky from a makeshift cannon in Colorado. That was on 20 August, 2005, which might hereafter be celebrated as 'Substance Abusers' Pride Day'. But the celebrants could not be expected to march well.

My friend Ralph was present in Colorado, far from his home in England, when my friend Hunter's granulated remains were dispersed by an explosion, and he surely should have been. Emotionally speaking, Ralph might as well have been saying farewell to a wife. That was how profound and dynamic his creative partnership with Hunter had been. Hunter's texts had inspired what were arguably the most passionately communicative illustrations Ralph had ever made. Yes, and they in turn had somehow made Hunter's texts seem even more entertaining and theatrical and exciting than they would have been without them. Let it be noted: like so many marriages of strong partners who are good for each other, Ralph's and Hunter's was a love–hate relationship. But, as Ralph takes pains to say in this good book, it was mostly love.

'Don't write, Ralph.
You'll bring shame on your family.'
Hunter S. Thompson

INTRODUCTION

A 150-foot monument is a tall thing, even in a majestic range of mountains in Colorado.

In daylight it shone with arrogance and the massive hope of a huge ego. It gathered its strength from some long-gone, youthful spirit that wanted to wake up the world – not all the world, but the tired New World of America, a country whose optimism had been dimmed by two world wars it never wanted to be part of – a couple of skirmishes it thought it could settle in the blink of an eye.

A wayward soul had dreamed of this strange monument back in 1977. An undertaker in West Hollywood had listened to the explicit demands of a man who knew exactly how he wanted to die and how his remains should be treated thereafter.

*

'I would feel real trapped in this life if I didn't know I could commit suicide at any time,' Hunter Stockton Thompson told me many years ago and I knew he meant it. It wasn't a case of if, but when. He didn't reckon he would make it beyond thirty anyway, so he lived it all in the fast lane. For him there was no first, second, third or top gear in a car – just over-drive. He was in a hurry.

On 19 February, a Sunday morning, I had just finished signing the twelve hundred limitation pages for a version of our 1981 book, *The Curse of Lono*, to be published by Taschen, which Hunter had signed so uncharacteristically – obedient and mechanical – during December 2004. I had thought this was very strange. Usually he had to be cajoled like a child to do anything like that. So I drew his portrait across the last sheet, his face glaring out, his two eyes in the two letter *o*s of *Lono*. I put the cigarette holder with the long Dunhill prodding upwards from his grimacing mouth, signed it with an extra flourish and closed the last of the four boxes.

The old bastard! He waited to make sure I had finished the task and then he signed himself off. I knew it had been too good to be true.

Now I would have to build that monstrous cannon in Woody Creek, a 150-foot-high column of steel tubes, with the big red fist on its top, his ashes placed in a fire-bomb in its palm. 'Two thumbs, Ralph! Don't forget the two thumbs!!' It was the Gonzo fist and he really believed I could do it!

He could be very persuasive. As a boy he was hired by the milkman to collect outstanding bills from the citizens of Louisville, Kentucky. He was shunned by his neighbours and especially by the town's literary establishment. So he had a score to settle.

Such were his demands as he tipped at his windmills. People were fucking with his beloved Constitution and he was born to banish the geeks who were doing it. In that way he was a real, live American. A pioneer, frontiersman, last of the cowboys, even a conservative redneck with a huge and raging mind, taking the easy way out and mythologizing himself at the same time.

I have always known that one day I would make this journey, but yesterday I did not know that it would be today. I had the good fortune to meet one of the great originals of American literature. Maybe he *is* the Mark Twain of the late twentieth century. Time will sort the bastard out and I leave it to others more qualified than me to assess and appraise his monumental literary legacy.

Ralph Steadman, Monday, 20 February 2005

THE 70s

'Ye Gods, Ralph! A matted-haired geek with string-warts!
They told me you were weird, but not that weird.'

THE KENTUCKY DERBY
May 1970

An innocent abroad & a meeting of twisted minds in
Bluegrass country . . . Eating out with Hunter . . .
Filthy habits & Mace gets in your eyes

Scanlan's magazine, for those of you who missed those nine wild months of publishing history, was the brainchild of Warren Hinckle III, who scorched through three-quarters of a million dollars of borrowed money in the pitiless pursuit of truth — not least the call to impeach Richard Nixon as early as 1970.

The magazine was named after a little-known Nottingham pig farmer called Scanlan and it dedicated itself to maverick journalism and anything that seemed like a good idea at the time. Warren set about making sure everyone knew everything about anything that moved in America, from covert activities in high places to rats in a New York restaurant kitchen. His business partner was Sydney E. Zion, who later gained a reputation as the man who fingered psychiatrist Daniel Ellsberg as the source of the 'Pentagon Papers', which had made public in *The New York Times* the US military's account of activities during the Vietnam War.

They achieved their goal and made Nixon's blacklist in record time. Unfortunately Warren's excessive lifestyle and appetites outstripped the financial cornucopia that was there to begin with. After the ninth issue, the well dried up and the magazine sucked itself to death. When it happened we were out on a limb, covering the America's Cup for them. Not the best news to learn over a bad line to New York while asking for more funds.

Scanlan's found me in Long Island in April 1970, not long after I had arrived from England to seek my particular vein of gold in the land of the screaming lifestyle. I was staying with a friend in the Hamptons to decompress. His name was Dan Rattiner and he ran – and, in fact, still runs – the local newspapers, *Dan's Papers* and *The East Hampton Other*. Dan was young and in love with Pam. Dan and Pam treated me with great kindness and hospitality but after a week I began to feel I was getting in the way. It was time to make my trip into New York to look for work. Dan had generously picked me up at the airport a week earlier. I roll my own cigarettes and, without thinking, I lit up in his car. Dan said, quite sweetly, I thought at the time, that they tended not to encourage such habits, particularly in a car, because it was a bit like 'giving cancer to your friends'. I gulped down the smoke. Then I lowered the window and choked the filthy excrement out into the city.

That was okay, even in 1970, and I respected his guarded request. It was then that I first saw the crossing sign at intersections which came up in green and red, pronouncing: 'WALK', and then: 'DON'T WALK'. I laughed about it for some reason. It was the tone. The command. The admonition. Whichever one you obeyed, you were guilty. I was already beginning to like the city. DRINK, DON'T DRINK. SMOKE, DON'T SMOKE. PUSH THAT OLD LADY OUT OF YOUR WAY. DON'T PUSH THAT OLD LADY OUT OF YOUR WAY. BOMB THE SHIT OUT OF SOMETHING. DON'T BOMB THE SHIT OUT OF SOMETHING. RULE THE WORLD. DON'T RULE THE WORLD. OKAY. FORGET IT. WE CAN DO ANYTHING. WHAT D'YA NEED? HAVE A NICE DAY! FOREIGN POLICY? WHAT WAS THAT?

It had kept me in a reverie until we got to the Hamptons. It was my first true vision of the American way of life – a slice of the American Dream. The law-abiding vision of madness contained in a mechanical device. It was the law masquerading as a road sign. DON'T, was the true mantra. Americans love DON'T. Thou shalt not. The bedrock of received knowledge – the Ten Commandments. The God-fearing pioneers who still had a long way to go. GO! DON'T GO. FUCK YOU GOD! We're on our way . . .

I spent the week with Dan and Pam enjoying their joy in

themselves and their genuine desire to be nice to strangers. It
was then that I began to think that it was time I moved on
and left them inside their euphoric bubble. It was time to go
into New York.

I was just about to leave, when I got a call from *Scanlan's*
Art Director, J. C. Suares. 'We bin lookin' all over for ya!' he
growled with a pronounced Brooklyn accent. 'How'd ya like
to go to de Kentucky Duurby?'

'The what?' I replied.

'De *Duuurby*,' he repeated.

'Oh, you mean the *Daaarby*!'

'Okay.' He said, 'De *Daaarby*!' We were in agreement.

'Anyways, how'd ya like to go to de Kentucky Duurby wid
an ex-Hell's Angel who just shaved his head, huh? and cover
de race. His name is Hunter Thompson.'

'Johnson? Never heard of him,' I replied. 'What's he do?
Does he write?'

'Sort of,' JC replied. 'He wants an artist to nail the decadent, depraved faces of the local establishment who meet there
every year fer de Duurby. But he doesn't want a photographer.
He wants sometink weird and we've seen yer work, man!'

Undaunted by the credentials and in complete innocence – for I am an exceedingly trusting man – I readily agreed to go. I packed my bags, and arrived in New York at *Scanlan's* 42nd Street offices, which were conveniently situated above a cosy bar serving Irish Guinness and flanked on either side by dark doorways, harbouring drunks.

J. C. Suares greeted me with some caution, as I recall, treating me like a hired hitman with a reputation which had arrived ahead of him. It was some time later that I learned that I had not been the first, get-Steadman-at-any-cost choice. Hunter had originally suggested Pat Oliphant of the *Denver Post*, whom he'd got to know locally. Oliphant, as it happened, was off to London to attend a cartoonists' convention and had declined the invitation to be Hunter's sidekick. I have to thank him or hate him for that, but he did save my first trip to America from being a total washout.

I was introduced to the editor, Donald Goddard, a kindly, shrewd man and an ex-foreign editor for the *New York Times*, who had picked up a book of my collected cartoons in England called *Still Life with Raspberry*, the very week I left for America. Don explained, in a little more detail and with reserved reassurances, how interesting this job might prove to be. Being an Englishman himself, he put my natural anxiety at ease as only another Englishman can who is far from home but armed with foreknowledge. I, of course, am Welsh.

On the way to the airport I stopped off at Don's apartment, where I met his wife, Natalie, a representative for Revlon, which was fortunate. You see, I had left my inks and colours in the taxi and was – as far as an artist is concerned, anyway – naked. Miraculously, Natalie had dozens of samples of Revlon lipstick and make-up preparations which solved the problem at a stroke. They were the ultimate in assimilated flesh colour and, bizarrely, those Revlon samples were the birth of Gonzo art.

I was off, and Bluegrass country was only a couple of hours away.

Finding Hunter, or indeed, anyone who is not a *bona fide*, registered journalist covering the prestigious Kentucky Derby, is no easy matter and trying to explain my reasons for being there proved even more difficult, especially as I was under the impression that this was a *bona fide* trip anyway and I was a properly accredited press man. Why shouldn't I think that? I assumed *Scanlan's* was an established magazine. As it turned out, *Scanlan's* had got me a hotel room cheap at a jerry-built complex called Browns!

From there on in I was on my own. Innocence and a Welsh way of asking directions, coupled with a look of utter bewilderment, stood me in good stead. I noticed this early on and acquired a knack of looking more bewildered, more innocent and more Welsh if things got hairy.

Not able to locate Hunter at the hotel, though he had, in fact, booked in, I decided to take myself off to the track, eager to see for myself the colour and excitement I had been led to expect. I carefully selected a sketch book, a couple of handy, felt-tip pens and a spy camera. I was imagining, in my naïve way, something like a New Orleans jazz carnival or a set from *Carousel*. My first impression, therefore, gave me a shock. Ugly people jockeyed and jostled for positions in an uncertain queue like a soup-kitchen line from the 1930s. I became a piece of worthless flotsam and I sensed my foreign politeness cut no ice here.

'I'm looking for the press room,' I said.

'Go buy a ticket,' snarled the cashier through the ticket office window. 'Dis ain't Ascot, buddy!'

'You borderline creep!' I screamed and slammed my dollar down with righteous indignation. It did the trick.

I passed through a green, corrugated portal into a tunnel which led to the centre-field where people with absolutely no important mission to fulfil sloshed around in a sea of empty beer cans, hot dog stands and obsolete form cards. Some people were camped out and well-ensconced with all the mod cons necessary for a comfortable three-day stint. This was carnival-centre and not an ounce of influence operated here. These were the Christians waiting to face the lions and the Romans were up there in the surrounding grandstands, making bets.

I retreated back the way I had come, looking around the

back of the seat scaffolds and finally located the stairs leading to the press box. Again, with a high charge of Welsh bewilderment, I talked myself inside, past a pleasant lady with a Southern drawl and got myself a beer at the bar.

From where I stood I could see the course and the finishing posts. Press men were typing away and phoning editors and commentating through mikes to their radio stations. The races were in full swing and I felt heavy with a sense of inadequacy. I couldn't type, I had no one to phone anything to, I knew nothing about racing and I couldn't even locate the one man who could fill me in and make me feel that I was here for a purpose. It is a feeling I have experienced on many subsequent occasions where I have been shot into the middle of some strange place by some magazine or other which believes that all credentials are bullshit and that the mere mention of its name will send officials into paroxysms of reverence and respect. It's understandable, as any magazine proprietor worth his salt will explain that the world is waiting with bated breath for his next issue. He must think like that or go under with plummeting circulation figures.

I had been watching someone chalk racing results on a blackboard while I finished my beer and was about to turn to get another when a voice like no other I had ever heard before cut into my thoughts, sinking its teeth into my brain. It was a cross between a slurred karate chop and gritty molasses.

'Er . . . um, scuse me, er . . . you . . . er . . . you wouldn't be from England, would you! Er . . . an artist, maybe . . . er . . . what the . . . !'

I turned around and two eyes firmly socketed inside a bullet head were staring at the funny beard I was wearing on the end of my chin.

'Er sorry . . . I . . . er . . .' The mouth, firmly wedged between two pieces of solid jawbone, hardly moved. 'I . . . er . . . thought you might be . . . er . . . !'

'I *am* from England,' I replied. 'My name's Ralph Steadman – you must be Hunter Thompson.'

'That's right. Where you been? I was beginning to worry. I thought you may've been picked up or something.'

'Picked up?' I didn't quite understand.

'Yes . . . er, police here are pretty keen. They tend to take an interest in anything different. The beard . . . er . . . not many of them around these parts.'

'Er . . . why don't we grab a beer and maybe talk things over.'

I was beginning to take in the man's appearance. He was certainly not what I had been expecting. No time-worn leather, shining with old sump oil. No manic tattoo across a bare upper arm and certainly no hint of menace. No, this man had an impressive head cut from one piece of bone, the top part covered down to the eyes by a flimsy, white tight-brimmed sun hat. The top half of his body was draped in a hunting jacket of multi-coloured patchwork and his bottom half wore slightly small, blue seersucker pants. The torso was prevented from crashing to the floor by a pair of huge, white plimsolls with fine red trim around the bulkheads. But they did seem in proportion as a foundation for what looked like a lot to support. Damn near six foot six of solid bone and meat, holding a beaten-up leather bag in one hand and a cigarette between the arthritic fingers of the other. His eyes gave away

nothing of what he thought he was looking at in me — a 'matted-haired geek with string-warts', as I found out later. Writers have a compulsion to tell all eventually, particularly journalistic ones whose only real reason for being a journalist anyway is to blast out the secrets they are entrusted with 'off the record' and surprise the world — or their editors.

'Maybe we should watch a race or two . . . get the feel of things. I used to live in Louisville, so this is my stomping ground. I could fill you in a bit,' he suggested.

We took a seat directly overlooking the race track and I relaxed with a beer in one hand and a sketchpad on my knee. Hunter had a notebook and made sporadic scribblings in red ink.

'Do you bet?' he asked.

'Well, I put two shillings on Early Mist in the Grand National in England back in 1953 and won at ten to one.'

'Mmm, maybe you should try again. You sound lucky!'

A race had just begun, so we chose a horse just for fun, to see how we made out, without spending a cent. The horse won, so we decided to try again for real.

But my luck ran out and a modest amount of *Scanlan's* expenses had disappeared within the hour. During that time we had sounded each other out and overcome our first impressions. We were getting along fine as he kept pointing out faces that, for him, represented the real Kentucky face.

'That's what we are here for,' he would say. 'Nail that and you have it.'

The last races were being run and there wasn't much else to do around the track that would be helpful, so we decided to get back to the hotel, clean up and go into Louisville to eat.

I had so far made no sketches, or notes, feeling far too intimidated to do either. But my head was buzzing with strange impressions.

Hunter had hired a bright red whale of a car and had stored two buckets of beer, on ice, behind the front seats. We stopped off at a liquor store and bought a bottle of Wild Turkey bourbon — a drink I had not been familiar with up to that point. It tasted good and went down even better, though compared to a good malt whisky it's still a clumsy way to get drunk.

I was beginning to settle in.

'Maybe we'll just get ourselves arranged and then we could meet my brother Davison in Louisville – he's expecting us.'

'Okay by me,' I said. I was busy watching him drive. From the very first moment, I could tell he could handle a car with consummate skill. He was the sort of driver who could never be a passenger. One hand held the wheel, the other his cigarette holder and a beer can. Between his legs, or resting on the seat, he kept a tall glass full of ice and whiskey. His consumption of each was carried out in nervous progressions. The cigarette holder, with lighted cigarette, was placed in the mouth, drawn on, taken out, and then, with the same hand holding the beer can and cigarette holder, beer was swilled. For a moment, the other hand came off the wheel which was held briefly with the free hand, the whiskey was swigged, put back down and then the driving hand was returned to the wheel. All done while turning corners or overtaking in the fast lane.

The alternative is that I would hold the wheel while he did all these things in quick succession before taking over again. When taking corners, the driving hand swivelled the wheel in one motion to a magic point on its circumference and the car followed the direction of the front wheels, whatever the speed. It had to. Some divine intervention – or Wild Turkey – kept it on the road. Very occasionally I would hold the wheel at the magic point and the car would do as it was told. The clutch, brake, accelerator and Hunter's deft footwork did the rest.

We needed to eat. I was in a slightly befuddled state by this time and the potent combination of watery beer and whiskey was bringing on a severe attack of drawing, as always happens when I start seeing unusual faces through a haze of controlled drinking. My body becomes a protective casing and lets me observe through the two keyholes on the front of my head.

Hunter's brother, Davison, was another big man. They breed them big in Kentucky even though they often seem to have small mothers. He was darker than Hunter, much darker, which prompted a drawing from me. We were sitting in a restaurant, Hunter and me on one side of the table and Davison and his wife on the other. Not a very kind drawing, as I recall, but it was only fun, anyway. However, you could see he was visibly shocked and it took me quite a while — many drawings later, in fact — to realize that Kentuckians and many other Americans for that matter, take things like that as a personal comment and even, in some cases, an insult, comparable to a smack in the mouth. Hunter was shocked and then horrified as I persisted in adding to the drawing, making it darker and darker and more hideous as lines covered lines. It is at a point like this that the drawing can take over and I am completely immersed in its development. It is no longer a sketch or merely a personal insult, but a battle between the drawing and my desire to mould and twist it. The drawing becomes more important than the subject. Other people were watching now. It was no longer a private affair. Hunter fidgeted and made lame excuses for my preoccupation.

'It's a habit of his. This is nothing compared to what he did at the track this afternoon. Hideous things. We had to Mace people to escape. That's enough, Ralph. Stop it! Must be the Wild Turkey! He can't hold his liquor! Sorry about this!'

I was smiling as I stabbed one last violent stroke across the page and signed it with a flourish. I handed it over. My subject had become a victim and his dark features blackened, resembling my drawing even more.

'My God, it's terrible! Sorry about this!'

Hunter was fingering the Mace gun he had with him. He glanced around furtively. His brother's face looked a little hurt and even annoyed. He didn't say anything but he must have been restraining an impulse to retaliate in a more physical way. Analysing his thought process, ten years on, I would say that his frustration mounted as he tried to come to terms with this violation and this new friend of Hunter's who could do such a thing with such a benign smile on his face. In normal circumstances he would have hit out, but here nobody was touching him. Yet something was bugging him. Something primeval that he couldn't quite grasp. I was still unaware of the horrendous insult that this represented to these people and I had begun to draw the waiter.

'I think we ought to go. Come on, Ralph, cut this out or I won't take responsibility for what happens!'

'But I'm just beginning to enjoy myself,' I replied.

'Well you won't, five minutes from now, unless you stop that filthy habit. In fact, let's go! This place is getting ugly.'

'But we haven't eaten.'

'We're not likely to either. Let's go! Look out!'

I remember the waiter bearing down on our table and Hunter on his feet; a black tube and a fine hissing sound. My eyes began to sting violently and I stumbled up, grabbing my sketch pad. I remember eyes staring from all directions, from dark corners of the restaurant, as we made for the door to the street. The fresh air hit me and eased the pain in my eyes and on my skin.

'Crazy fool!' Hunter screamed. I don't remember any more.

I awoke on a bed, in my hotel room. My head throbbed like a train's bumpers shunting trucks in a railway goods yard. There were pieces of sketchpad strewn around, covered in drunken scrawls of half-formed faces I vaguely remembered. I reached over and managed to grab one or two and gazed at them. Did we eat? How did we get back? What have I done? Why am I here? The drawings told me everything. I was out of control. They were the scribbles of some raving drunk. This would not do at all.

My eyes were sore and my skin felt as if it was melting. I took a shower and massaged my head with the tepid spray to convince my body that I was just waking up and, in a while, my eyes would shine with vigour and my cheeks would glow with health. An old Boy Scout trick. Throw off the day before, as though it never happened. Everyday I am reborn. I felt better and tried to salvage the pieces of paper, as though they were vital pieces of information for my assignment. Who knows, one of those scrawls might contain the essence of a Kentuckian face. I keep everything — every piece of paper — or I sell it. But I keep it more often and if I find it by chance, I believe it to contain imagined properties which crave to be seen.

I have always found that whatever time I get to bed, even if it's six in the morning, I'll be up at eight, unable to succumb to the slovenly habit of lying in bed to catch up on sleep. I will feel hellish until the following night, when I can go to bed again at a decent hour and pretend my pattern is as regular as ever. It is partly my behaviour as a creature of habit and surely is the result of my small Welsh town, church upbringing as well as that Boy Scout training. Naturally, continuing this practice while away from home helps me to keep myself together, maintain a respectable front and sustain what would otherwise be a complete breakdown of my fine Welsh constitution.

Hunter was a different animal. He seemed to gain strength from rakish marathons. I am certain he learned the secret of maintaining a drug-racked body from an old Indian in the Appalachian mountains. He learned the balance between living out on the edge of lunacy and apparently normal discourse with everyday events. Whatever reaction he adopted towards a situation, whether it was giving a hell-raiser speech from the interior balconies of the Hyatt Regency Hotel in San Francisco or firing a Magnum .44 at random into the night in front of strangers, he would always convince those around him that they were the ones who were mad, irrational or just plain dumb and he was behaving as a decent law-abiding citizen.

The hotel was one of those condominium-style, broadly laid out, jerry-built establishments, the sort of place designed to cope with mass suicides. Hunter's room was on the other side of what could loosely be called a courtyard.

Today was Derby Day. I had taken breakfast and thought that, as it was now about nine-thirty, it would be a good plan to wake my friend, who must have managed to get that extra hour's sleep.

I banged on his door and waited. Nothing. I banged again more emphatically and thought I heard a muffled cry from somewhere inside, like the sound a whale makes when searching for its mate. I banged again and called his name. A cry like twelve rhinos in mortal combat shook the flimsy walls and doors.

'Why torment me!'

'It's me, Hunter. Ralph.'

'No! I need sleep.'

'Open the door.'

'Why? Goddamn it!'

'Don't you think we should be making our way to the track? It's Derby Day.'

'Crazy English faggot. Go back where you belong. You faggots are all the same.' The door remained closed. I ignored the eccentric insults and tried again.

'Just open the door and we can discuss it.'

'Come back in another two hours.'

'Okay, okay. I'll go and observe people in the lobby.'

A mumble suggested that our conversation was at an end.

I observed for one hour with horrified fascination! Obesity was a common sight in most central and Southern states. The culmination of years of hash browns and eggs sunny-side up, French fries, oily salads, pancakes with maple syrup and club sandwiches, washed down with sweet Coca Cola and blueberry milk shakes.

It is no accident that men's trousers are made in stretch-nylon synthetics, to accommodate their drooping paunches and elephantine legs and women's dresses hang like marquees without tent pegs and guy ropes.

The swamp-like humidity of a place like Kentucky brings rivers of sweat pouring over these undulating plains of wobbling flesh, dripping onto polished vinyl floors or down their socks into maroon and white sneakers. I spent a while at the counter of the breakfast shop talking over one or two black coffees.

A lady dressed for the races in a mint-julep-green, two-piece trouser suit and white, simulated tortoise-shell shades with extended frames like bridge spans sauntered up to the counter.

'Two hamboigers, please.'

The server responded with a slovenly turn away towards the hamburger cook at the cooking counter.

'And have you got any relish?'

I hastened to offer her the tomato ketchup standing near my elbow on the counter, thinking it was what she required.

'Dat's not relish, stoopid!' she sneered.

'Oh, I beg your pardon,' I replied. 'I'm not familiar with these things. I'm not sure I know the difference.'

'Well, ya oughta,' came the final response and she turned away and never looked my way again.

I made my way back to try Hunter again. Using the same approach, I banged fiercely on the door. I got the same response and a worse stream of abuse than the first time, but this time I wasn't giving up. I had to show the bastard that we Welshmen don't take this sort of thing lying down. The door flew open, but there was no one there. Peering through the entrance I caught sight of a bare backside disappearing back beneath the sheets.

'I really think we ought to make a move, Hunter. It's getting on for midday.'

I pulled the curtain open and light flooded in. A squinting dome of a head peered over the sheets, eyes contorted into scars as the fierce light bleached his features into a crumpled carnival mask. It was a ghastly sight.

'Er, um, my God, what time is it?'

'Midday' I said.

'Goddam, why didn't you wake me. We should be at the track.'

'I did, but you wouldn't get up.'

'Hmmmm.'

One hour later we were sitting in a roadside coffee shop-restaurant. Hunter had drunk a litre of orange juice, consumed about thirty different kinds of vitamin pills, two grapefruits, two club sandwiches, four Bloody Marys and was just on his second Heineken with a scotch on the side. I have since come to suspect that he was a secret health freak who worked hard at it, devouring great gobs of health food and goat's milk in vast quantities to build up his flailing insides to concert pitch before any bout with humanity.

'Got to stoke up for the day ahead. It could get nasty. Aren't you going to eat?'

'Well, I did actually,' I replied. 'I had a lightly boiled egg and half a grapefruit a couple of hours ago. I'm fine.'

'Ye Gods! You won't make it. This could be a day of ugliness. Horrible, horrible! People drinking and eating like pigs and vomiting everywhere. We'll be sliding in the stuff before the week is out. And keep that damn thing out of sight.'

I was idly sketching a few of the late breakfast eaters as he talked.

'Sorry,' I said, 'but I did come here to observe and draw.'

'Yes, but why must you scribble these filthy ravings and in broad daylight too?'

'You must remember, Hunter, I am seeing everything for the first time and it's a culture shock that is affecting me strongly. Anyway, they're only drawings.'

'Don't even think that here, Ralph,' he replied, looking around furtively. 'These people are primitive head-hunters. You're liable to end up in a ditch, kicked to death by a horse.'

I had begun to notice amongst all these stern warnings a hint of a twinkle in those tight eyes of his, as though he was secretly enjoying the possibilities that such a strange practice as mine might provoke. He had been used to working with photographers on other assignments and the detachment with which a photographer usually works gave him nothing against which he could spark. Here was something that could conceivably become a part of any story he worked on. An ongoing situation that could rear up and enter the scene at any unexpected moment, though I did not realize this at the time.

I was also looking at a worried man. There was something on his mind that he had not yet told me. He had no stomach for this story but he was going through the motions of appearing to be working on it. He had seen it all before and worse; he had been writing stories like these about sporting events and Chicago riots *ad nauseam* and right now it all seemed like a pointless way to go through life. What he had not yet let me know was that he had harboured a project since very early on that would help him to burn his own particular hole in life, but, in the process, would burn him too by the time he was forty.

But here he was, large as life, at age thirty-two, back in his home town, trying to crank up enthusiasm for some rich old vein of resentment he had felt a long time ago while still a youth, a reject from the powers that ran this God-awful place, and he was saddled with some weirdly bearded innocent abroad who was naïve beyond belief and who seemed unable to comprehend that any activity emitting hostile waves was bound to be interpreted as a fair provocation for physical violence by these people.

But maybe that was the reason for the twinkle in his eye. It was likely that at some point I would create some ghastly confrontation with an insulted cattle-rustler who would loom bigger than either of us, even if we both wore cowboy boots. That would be enough to start the juices flowing in anyone with a hint of journalistic flair. Perhaps the light of Gonzo was already beginning to streak across his tarnished mind while I was innocently attempting to contribute something tangible to this assignment. I doubt that it was a fully formed idea from the very start, most good ideas rarely are; but this was surely the stuff of which stories can be made.

The Kentucky Derby alone was certainly no reason to be here. It had been written about annually by armies of reporters since it had begun, but to find himself back on home ground with only a record of disillusionment in his soul, no prospects and an unfulfilled wish to have snuffed it at thirty, there had to be something else. If you add to this the fact that at that very time he was experiencing severe family problems too — having to have your mother placed in an institution of care is

severe. The stage was set for a weird chain of creative responses in the mind of anyone on that particular high wire. This was no ordinary homecoming. This was a do-or-die attempt to lay the ghost of years of rejection from the horse-rearing elite and the literati who sat in those privileged boxes overlooking the track and the unprivileged craven hordes who grovelled around the centre-field where he had suffered as a boy.

Even the horses got better treatment, particularly if they showed signs of becoming champions, though they too were shot within ten minutes if they broke a leg.

'C'mon, let's go. We'd better get over there. You've got to see inside that clubhouse before the big race. Maybe you'll see then just what you're up against. *Scanlan's* could get no passes so we'll have to talk our way in.'

At the time, talking our way in didn't seem as much of a problem as Hunter made out but maybe that was eased by the number of people he seemed to know around the Privilege Stand. Real, blue-glass rich, young who had either inherited their hoard or started a second-hand car-mart ten years earlier. They were extremely friendly and had been made pliant by the mint juleps, another drink I had only just discovered. They had their own fridge-full under their seats. With their help we acquired the necessary 'hail-fellow-hi-there – anyone who's a friend of Jake's is a friend of mine' treatment.

'Hell, fella, you from England? Me and my lady did London back in '66. Say, those bobbies of yours sure know how to be kinda friendly. We sure would love to go back someday. Mebbe when you get rid o' that commie government of yours. You guys deserve sompin betta.' And so on.

This was May 1970 – and less than two months later, that desire was fulfilled when we elected a great big smiling Jack of a PM called Edward Heath.

I remember one particular young Kentucky belle, who was warm towards me to such an extent that Hunter was whispering over my shoulder to be careful as it was the wife of a friend of his. Unfortunately, our mutual attraction was the cause of a nasty scene later that night, when I drew her with unpleasant results. However, by that time I was beyond the stage of gentle decorum and was reacting violently to the sights and sounds around me. I drew anything in sight and wasn't particular who saw the result.

The clubhouse, as far as I remember, was worse, much worse than I had expected. It was a mess. This was supposed to be a smart, horsy clubhouse, oozing with money and gentry, but what I saw had me skulking in corners. It was worse than the night I spent on Skid Row a month later, back in New York. My feet crunched broken glass on the floor. There seemed no difference between a telephone booth and a urinal; both were being used for the same purpose. Foul messages were scrawled in human excrement on the walls and bull-necked men, in what had once been white, but now were smeared and stained, seersucker suits, were doing awful things to younger but equally depraved men around every dark corner. The place reminded me of a cowshed that hadn't been cleaned for fifteen years. Resolutely, I stayed there, but really only because of my own desperately befuddled and drunken state. Somehow I knew I had to look and observe. It was my job. What was I being paid for? I was lucky to be here. Lots of people would give their drawing arm to be able to see the actual Kentucky Derby which was now hardly an hour away. Hunter understood and was watching me as much as he was watching the scene before us. We stumbled around like a couple of lapsed Catholics searching for something to believe in.

A bottle hit the wall and its contents spewed down in a bubbling stream to the floor. It is strange how one seems to gain drunkard's immunity from personal injury when completely plastered and by now we had swilled more than we cared to think about. Green juleps and amber fluid heaved about in our stomachs like oil-slicked waters in a harbour storm.

Something spattered the page I was drawing on and, as I moved to wipe it away, I realized too late that it was somebody's vomit. A ghastly-looking gargoyle-faced horse-dealer had hurled from about five feet away and the pressure-driven contents of his insides had miraculously missed me and caught the elevated plane of my sketchpad sufficiently to splatter down the page in an array of colour which would have been a wonderful effect artistically, if it had not been for the evil smell.

During the worst days of the Weimar Republic, when Hitler was rising faster than a bull on heat, George Grosz, the savage satirical painter, had used human shit as a violent method of colouring his drawings. It is a shade of brown like no other and its use makes an ultimate statement about the subject. Its organic nature lends a powerful presence to the image and no one misses the point.

'Seen enough?' asked Hunter, pushing me hastily towards an exit that led out to the club enclosure. I needed another drink. So, we made our way back to the Privilege Stand to wash ourselves down with a few more mint juleps. The young lady whose fancy I had taken was still there, drawling on in her Southern accent.

'Er . . . one more trip to the inner-field Ralph, I think,' I heard Hunter say nervously. The young lady's husband was eyeing me and pretending to be friendly, presumably to hide what other thoughts he may have had.

'Only another half-hour to the big race. If we don't catch the inner-field now, we'll miss it.' So, we went.

Black faces talking bets. Tripping over hot dogs half-eaten in piles of Coke and beer cans, bodies in a final state of consciousness before dying, or just sleeping. Ripped T-shirts and odd-shaped hats, home-painted to say their piece: polit-ical or just plain saucy.

While the scene was as wild here as it had been in the club house, it had a warmer, more human face, more colour and happiness and gay abandon − the difference in atmosphere between Hogarth's *Gin Lane* and *Beer Street*. One harrowed and death-like, the other bloated with booze but animal-healthy.

It seemed to me that we could just make it back before the race began. We had placed our bets, so we fought our way back to our friends in the stand. They plied us with more juleps, waited and rolled about. I had chosen some religious-sounding name. Later, when I looked up Hunter's Kentucky piece I saw it had been called Holy Land.

It was then that I conceived the idea to turn and face the

crowd the moment the race started. So what, I thought, I hadn't come to see the race anyhow. It was the people and what better than to observe them utterly engrossed in their hopes out there on the track? What stories their faces would tell. I would photograph them constantly throughout the race and get myself a group of pictures that would surely reveal the true Kentucky face.

Nobody noticed me facing the wrong way, but it was an unnerving experience. I was experiencing the full weight of an actor's-eye view of this energized audience whose emotions individually and collectively pulsed forward towards the track as the moment arrived, one second before the race.

I stood my ground, though my impulse was to turn the same way as everyone else. 'Hey boy! You're facing the wrong way, boy!'

Now, the last time I was called 'boy' was the day I decided to grow something on my chin. I was sick of the 'boy' reference. I could hold my own with the best of them and the first time I heard it, I was all of twenty-three and newly married. So my beard represented a sign of my maturity, rather than mere decoration. Now I was thirty-three and this Gadarene mass of depravity was saying the same thing, not to mention that I also have prematurely white hair from an overactive pituitary gland.

This warranted more than a camera shot. It was drawing time again and out came my sketchpad.

'Hey, Steadman! Hold it!'

Hunter was already anticipating trouble.

'But he called me "boy"!' I protested.

'Leave it! In this town they call everyone under eighty "boy".'

'I still don't like his tone.'

'You'll like his anger even less! So put your book away.'

'Hmmm.' I succumbed to my natural fear of physical violence and put my sketchbook away.

'I'll take his picture instead and do the bugger later. He's the type I think you are looking for — the Kentucky face.'

'I'm looking for something worse than that. The nearest you've come to it so far was the picture of my brother and

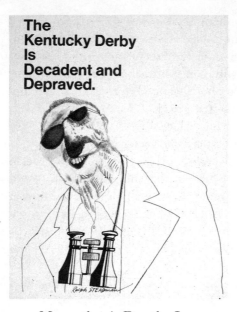

The Kentucky Derby Is Decadent and Depraved.

that worries me. My mother is Dorothy Lamour compared to that twisted pig-fucker you were about to tangle with.'

I was watching the man's binoculars dangle around his belly.

'Okay, later,' I said. 'Hold it, sir! Photographer for the *Courier, Journal and Times*. Wanna be in *pre-int*, dontcha?' My accent wasn't bad considering I'd only been in the territory a couple of days. 'Just tell me if I win, ole buddy bud, and howsabout a smile?'

'Right on, fre-yind,' and a 22-carat gash appeared somewhere around the place where his mouth was supposed to be. It was like the back end of a goat with its tail up.

'Thank you kindly, sir.'

Snap!

'Got you, you bastard!' I thought, keeping it to myself.

The other spectators were too engrossed in the race to bother about some drunken photographer snapping like a demented tourist. It was good publicity, anyway. Might end up on the *Louisville Courier* front page – friends of the winning horse.

Another asshole!

Who would have thought that I was after the gristle, the

26

blood-throbbing veins, poisoned exquisitely by endless self-indulgence, mint juleps and bourbon? Hide, anyway, behind the dark shades, you predatory piece of raw blubber. Is that your wife by your side or a cake of wrinkled make-up, waiting for its owner to return. It was moving with heaving animation, glowing with rouge – an ocean of rolling surf breaking on wet sand, cutting sharp moments of wrinkled sweat into an undulating mask beneath cantilevered shades that sparkled with diamante rims. Some of the pink dinosaurs were slightly younger, softer versions, daughters cleft from pedigree stock to preserve the line. Interbreeding was rife, no doubt, and new blood would only be injected by healthy young rednecks whose ticket to these inner circles came through the stables where they would eventually become trainers and lay an owner's daughter as soon as was decently possible. The older versions wore light, bright, summer coats, either plain white with acid, julep-green edging or outrageous sack-shaped coats in checks as broad as a freeway.

The race was now getting a frenzied response as Dust Commander began to make the running. Bangles and jewels rattled on suntanned, wobbling flesh and even the pillar men in suits were now on tip-toe, creased skin under double chins stretched to the limit, into long furrows that curved down into tight collars.

Mouths opened and closed and veins pulsed in unison on stretched necks as the frenzy reached its climax. Sweat rolled down the flesh-furrows into their collars. One or two slumped back as their horses failed, but the mass hysteria rose to a final orgasmic shriek, at last bubbling over into whoops of joy, hugging and back-slapping. I turned to face the track again, but it was all over. That was it. The 1970 Kentucky Derby won by Dust Commander with a lead of five lengths – the biggest winning margin since 1946 when the Triple Crown Champion, Assault, won the Derby by eight lengths.

Already people were making a beeline for the exits. There was nothing after that except a few minor races for the diehards and bet-aholics and the presentation of the Derby trophy, but I could get all that from the newspaper. I had watched the Derby through the behaviour and wild responses of the crowd.

I was probably the only person on the track who had done so and it separated me from the general activity.

The only thing to do now was to collect all the follow-up papers I could lay my hands on – winning trainers, their ladies, jockeys, champagne bottles and owners – have my first night in town and get out with my freshly stuck vision of it all. But, I also had to catch that Kentucky face if it was at all possible – particularly that man who called me 'boy'!

Something that came back to me as I tried to recollect exactly what happened that day was the sight of a beautiful copper-beech brown horse that had broken its leg. You could see it was broken by the angle it tilted from the ground because of the pressure brought to bear on it as two men tried to pull it along towards a waiting horse-box where it would be driven off out of sight to be shot. I can still hear its screams but I seemed to be the only one who noticed. It was as though it was a regular occurrence and it probably was, although for me it was an act of the most savage brutality. I guess familiarity numbs the senses when you've been around horses long enough.

As the crowd dispersed, Hunter suggested that we should spend a little more time at the press box, where we could top up on a few more whiskey and beer combinations. I'd had enough juleps and I needed something to turn my stomach back from green to its natural colour of lumber-brown. We must have been blind-drunk for I don't remember the drive back to Louisville, where Hunter suggested we call into a club called the Pendennis.

I vaguely remember getting out of the car and, as if by magic, the girl with the Southern drawl was climbing out of one side of a sleek, low two-seater parked beside us and her man was clambering out of the other. I was feeling pleasant now – and drunk enough to ignore her husband.

The club could have been transplanted in spirit from a London club, except for the marvellous atmosphere that, I imagine, one would find in the hallway of those rich planta-tion owners' mansions with white Palladian porticos and black flunkies standing at attention near every pillar, waiting to serve. We stumbled through to the drinks room – at least, I presume it was the drinks room. Drinks appeared miraculously wherever I put my right hand. There were pool tables in the room and, fancying my luck with the Southern beauty, I challenged her to a game.

'Ah cain't play, but you can teach mer.'

'Fine. I play a lot in England,' I lied.

Actually, the last time I'd played had been at the age of sixteen in a Welsh golf club. I never was one for games, except at moments like this.

My eye must have been particularly acute at that time in spite of the drink, or because of it, as, showing off in front of this lovely young thing, I sank every ball and left her full of helpless admiration.

The rest of the party was sitting in easy chairs, half-watching, but mainly chatting over old times. We joined them and I picked up my sketchpad again to draw the young lady.

'Steady, Ralph, you're drunk.' Hunter started to worry again. 'We don't want trouble here. You'll get us thrown out.'

But she seemed to be fascinated and so I set to work drawing in my freest style.

I ripped the page out of the sketchpad and handed it to her. She smiled as she took it but as her eyes struggled to decipher my lines, a horrified look crept over her face. Meanwhile, her smile fought to stay there but began to turn instead into a twisted grimace.

'That ain't purty . . . I'm purty, ain't I?' she blurted.

'You're beautiful,' I said, 'but this is just a fun drawing. Don't take it to heart. Oh, I *am* sorry, I really didn't mean to offend you. Here, let me try again.' And I started another. It somehow got worse than the first and the atmosphere around me became quiet and fascinated, though weirdly cool.

'Wha'd'ya draw me lark theyat?' she drawled. 'Ain't ah purty? Ah'll draw you! Le-yen me yower pe-an and yower pey-per.'

She grabbed the drawings and began to hack at the paper, desperately trying to twist the lines into some awesome image of me in a futile attempt to get her own back. I had beaten her at pool and then it was as though I were now mocking her with a drawing. I knew then that I must never do this again or, at least, keep it to myself.

She was mortally offended. Any assignation we might have planned now lay in ruins in the pathetic scribble she was fighting to bring to life.

Her husband was ready to do something about me, but Hunter intervened, stumbling up to suggest we leave for some prearranged schedule somewhere at the other end of town and all meet later.

The husband had his coat off but I didn't realize it was for my benefit. There were lots of asides.

'Let's get drunk,' I said for something to say.

'You *are* drunk, goddamn you, and helplessly out of control. That filthy habit is getting worse. You'd be better off doing needlepoint!'

It was a suggestion I remembered and turned it to good use later — hard-core political needlepoint during the Watergate trial, when my wife Anna became Madame Defarge by a Washington poolside while I worked on an assignment for *Rolling Stone* magazine.

'I think it's time I was thinking of getting back to New York,' I suggested next morning. 'Let's have a meal somewhere and I can phone the airline for plane times. What day is it? We seem to have lost a weekend.' I hugged myself, trying to stop shaking. 'I need a drink,' I said.

'You need lynching!' Hunter was not yet up but was making an effort to rouse himself. As always, I had awakened at eight on an unknown day and had gone through my Boy Scout routine to freshen up.

'You've upset my friends and I haven't written a goddamn word. I've been too busy looking after you.' He was still in bed. 'Your work is done. I can never come back here again. Why don't you go back where you came from and let me sleep. Go and plague someone else with your problems.'

We laid the drawings out two hours later as we ate capsized omelettes and drank stewed coffee. I was feeling bad now and Hunter was preoccupied, worried.

'This whole thing will probably finish me as a writer. I have no story.'

'Well I know we got a bit pissed and let things slip a bit but there's lots of colour. Lots happened.'

'Holy Shit! You scumbag! This is Kentucky, not Skid Row. I love these people. They are my friends and you treated them like scum.'

'Well, I'm sorry. I just tried to be objective. I'm here to find something worth publishing. It wouldn't do to go back with picturesque landscapes of Kentucky and elegant horses. You showed me the depravity. You must remember, I've suffered culture shock. It's not easy to stay normal. I'm sorry. I'm more worried about your story, because without it my drawings are worthless. Without the words they lack authenticity.'

'Fuck the story! At this point there is no story. I grew up here. I went away and I came back. You appeared on the scene with this stuff like a travelling priest peddling twisted morality.'

'I'm just trying to do my job.'

'You miserable Golom! That's what Hitler said. Shit!'

Hunter had brought his fist down on the omelette set out for him in the coffee bar we had gone to. I wiped feebly at the scrambled substance sticking to my already crumpled sports jacket in an attempt to lessen the shocking effect this would have on my alarming countenance. I had only been there a week.

The few patrons who, for masochistic reasons known only to themselves, had ventured into this World War II concrete tank-trap of a coffee shop were feverishly finishing their respective beverages as they fumbled for ten-dollar bills or anything that would amply cover their tabs without having to wait around to pay at the desk. The scene was getting ugly and ten dollars was, I expect, a cheap price to pay for the preservation of the clothes they were trying to sneak out in and, anyway, the cash till and the proprietor, himself, were covered in the stuff, so getting change would have been a messy business and, perhaps, a little weird.

I eyed the proprietor nervously, expecting him to reach for the phone – 'Hello, is dat the police? Some maniac has just sprayed my coffee shop with omelette. What kind of omelette? Well, there's tomato on the wall mirror and this is mushroom on the phone here and that looks like a piece of red pepper on da cash register. It's probably the Spanish omelette, officer . . . what . . . yeah . . . they're delicious. What time do we . . . ? Well, we usually stay open until midnight . . .'

It was only the excruciating stinging in my eyes that snapped me out of my daydream — a defence mechanism against a situation I couldn't handle. I heard Hunter raving on as I stumbled about helplessly trying to retrieve my drawings and baggage.

'I've wanted a worthy cause to try out this chemical billy properly on all week and now I've found it, you scum-sucking geek! That's it. Chew on that for a while.' I heard a quick hiss from the spray can Hunter was brandishing. He had Maced me again! I choked and ripped at my shirt in an attempt to stop the stuff burning my skin. My flailing hands caught something — a beer can — and I tilted the contents over my face and chest to relieve the suffocating pain. I screamed with relief but the pain returned immediately.

'Arrrrgh! Air, I need air! . . . outside!'

A hand gripped my coat and shoved me forward violently. The clammy heat of the midday sun intensified the burning and the hot tar of the car lot outside scalded the palms of my hands.

'My drawings — where are my drawings? I can't see!'

'I've got your drawings, you worthless faggot. Get in the car — you're causing a scene. Do you want to end up in jail too?'

There was no sympathy in Hunter's voice now. It was as cold and hard as a municipal sewer pipe. I crawled into the open door and grabbed wildly for the ice bucket I knew was behind Hunter's seat, splashing the still cool water over me with a cupped hand.

'Mind the seats, goddamn it. This is a rental car.'

The cool water eased the searing sting momentarily and I felt the car roar forward as Hunter angrily threw it onto the freeway.

'I'm taking you to the airport, though God knows why I

should try and save your wretched little ass. If you weren't a weird stranger I'd let them hurl you in jail but I don't need you on my conscience.'

'Thanks,' I said putting my head inside the bucket which was now on my knee.

'Water doesn't help either,' he said. 'Luckily you only got a mild dose. Apart from losing your eyelashes and three layers of skin you'll be right as rain when you reach New York, although explaining your appearance to airport police might be a little difficult.'

'Oh don't worry,' I said, 'my skin is allergic to the stifling humidity of tropical swamps like this one anyway. I'm surprised I haven't felt like this sooner.'

The front of my shirt was wet and reeked of vomit, but it couldn't be mine because I could not remember being sick. Hunter was driving angrily, cursing the traffic as well as me. I saw my face in the rear-view driving mirror. My stinging eyes looked like crushed pomegranates and I remember having a coughing fit. I was not well. Hunter skidded violently to a stop in front of the terminal. He reached over and opened my door for me and he threw such a barrage of verbal abuse at me, I knew I had really upset him. But I didn't know why, since I was the injured party. 'Now bug off, you worthless faggot!' he snarled. 'You twisted pigfucker!' He laughed maniacally at the sight of me. Then still half-laughing he gurgled: 'If I weren't so sick I'd kick your ass all the way to Bowling Green – you scum-sucking foreign geek. Mace is too good for you. We can do without your kind in Kentucky. Now get your bags and get out, and take your rotten drawings with you!' Then I watched him go, his wheels squealing on the tarmac. That was a successful trip, I thought. I fumbled to find my sunglasses to cover my eyes and walked up to the check-in desk.

I cleaned up in the aircraft toilet during the flight and on arrival in New York made my way straight to *Scanlan's* offices to lodge a complaint. However, the offices were closed and I took a beer in the bar downstairs while I wrote a note to accompany the drawings, explaining the oily stains from the omelette and the reek of the cosmetics I had used to colour them. I made my way back to the flat I was staying at in West 11th Street in Greenwich Village and collapsed onto the bed.

I was awoken from my deep torpor by the phone. It was Warren Hinckle III.

'Yeah?'

'Hey, Steadman, is that you?'

'Er . . . yes . . . did you get my stuff?'

'You're a messy worker Steadman – I like it. Let me ask you; didn't you see any horses?'

'Quite a few.'

'Den why ain't we got any?'

'Ah that, yes, well, I'm doing the horse one today.'

'Okay, let me see it. And what'd you do to Hunter? He says he can't write and he's blaming you for arresting his creative flow.'

'I'm sorry, he seemed to have personal problems to overcome. Home town, past memories, meeting old friends, mother, and so on. I'm sure he'll produce something soon.'

'Didn't you two hit it off or somptin?'

'Got on like a house on fire,' I said. 'I'll be along to the office just as soon as I finish this horse drawing.'

I rang off and stood silently for a minute or so, staring blankly at the shaking right hand I had raised to wipe the sweat from my brow. It was early May and the heat was beginning to build up in the city.

Before I left New York a month later, I asked Warren Hinckle III out of curiosity about my Kentucky drawings. I rarely part with my work and I certainly would not have willingly parted with those ones. I was assured by J. C. Suares that they had been taken by Warren, personally, to the printers in San Francisco on one of his twice-weekly east–west coast flights with Sidney Zion – 'for safety', he added. So I asked Warren where my drawings were.

'How the fuck should I know, Steadman?' was his brusque reply. 'Maybe someone wiped their ass on 'em!' he twinkled out of his one good eye.

I never did find out, but, wherever they are, they still belong to me . . .

THE AMERICA'S CUP
September 1970

*Back to America to live on a boat . . . Blood-red dye in
the water . . . Seasickness, paranoia & distress flares in
a dinghy in Newport, Rhode Island*

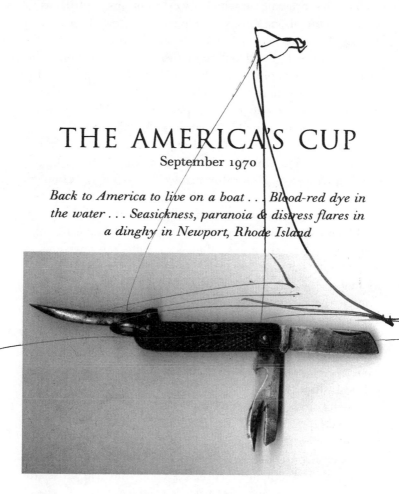

I had returned to England in the June of 1970, prompted by
an invitation from the London *Times* to cover the British
General Election — the one that would decide for ever that
Britain was, indeed, becoming a socialist state and that the
'white heat of technology' was burning a hole through
the consciousness of the population; that the old order of
national conservatism was a spent force and we would prevail
with our socialist ideals. Harold Wilson was top dog and
Edward Heath was the Opposition.

35

The first drawing on which I worked on my return was to fix my point of view. It was a picture of two large voters looking down upon three pedestals on which the three main contenders are fighting out their respective campaigns. The caption read: 'Happiness is a Small Politician' — my mantra then and forever more. On Monday, 8 June, I had decided that the two main contenders were indistinguishable from each other and they became Mr Weath and Mr Hilson.

I followed with a desert landscape showing a sun burning down on the figure of Edward Heath dying of thirst. In the sun was written 'A Better Tomorrow' and foolishly I thought that would nail him. It didn't, of course, because he won. But I had decided not to like any of them.

I followed that with another landscape called *Early Morning Scene* showing Heath in his pyjamas watching the sun come up with Harold Wilson's face on it. Another landscape followed on 18 June — Election Day, featuring Heath, Wilson and Jeremy Thorpe, the Liberal Party leader. Their faces made up the un-dulating details of the earth. I called it *The Wasteland*. The media was awash with these figures and their versions of a better future. I was against all of them in principle and my wholly negative stance represented a subjective approach to political cartooning. It was more intuitive than informed.

I don't think that at the time, or now, come to think of it, I gave a damn. Foolishly, I wanted truth and idealism, but there was none to be had. I felt early on that these people were merely electioneering and none of them stood for anything but them-selves. I had become an incurable sceptic who had no secure place on any newspaper, but I needed the work.

I persevered in that vein however and one particular cartoon from that period called *The Unhappy Clown* contained a kernel of truth. I dressed all the political figures of the time as clowns and depicted them beating each other over the head with balloons on sticks — except one: Enoch Powell. Nobody would play with him because he was standing there, holding a mace on a chain. He was, of course, the only one who wanted to play hardball. There were the troubles in Northern Ireland and the fast-emerging race issue and Powell wanted to face them both, as well as Rhodesia and South Africa. His 'rivers of blood' speech, some time later, marked him down as an embarrassment to the Tory party. He was, of course, saying what many simply thought, right or wrong, but he was hardly a vote-catcher. On 16 June I did a picture of Powell as a fly sitting on a heap of shit. Ian Paisley, a rabid Unionist, was beginning to make a lot of noise and he is shown flying in to alight on the same heap. Edward Heath creeps forward with a fly-swat to hit them both while Enoch is saying: 'Go find your own heap, Paisley!'

Assuming my work for *The Times* was finished, I started looking elsewhere, but I was immediately offered another trial period as a *Times* staff cartoonist for three months, which I accepted. Most of my work from that period was morose and uncompromising. I assumed that old age pensioners and the Yippies of Abby Hoffman and Jerry Rubin were on the same side – behind the barricades of fridges and TV sets challenging the Establishment and that was the side I was on. The editor at the time was William Rees-Mogg, who was continually warning me that my attitudes were dangerous and subversive, but the assistant editor, the late Charlie Douglas-Hume, was

on my side and I was quite unaware that my opinions were the source of some embarrassment at editorial meetings. I made editorial drawings to go with articles and did odd jobs for the *Sunday Times*, as well as the *Radio Times*, and I cut my ties with *Private Eye*.

I had moved into a borrowed flat at 40, Elm Park Gardens, Chelsea, with Anna, to whom I have now been married for thirty-three years, and attempted to hold down this uneasy job. I had four children from a previous marriage and needed to sustain outgoings of a serious nature. I hunkered down and set about earning a decent living. Anna unswervingly accepted my status as an artist and a man with a mission. Together we set about making something unsettled into a framework for a promising future.

We settled down for the rest of the summer.

I received a letter from Hunter dated 2 June 1970, the first since my return to England.

> Dear Ralph . . .
> You filthy twisted pervert. I'll beat your ass like a gong for that drawing you did of me. You bastard . . . stay out of Kentucky from now on. And Colorado too . . . Fuck you!
> And so much for that. I just saw the June Scanlan's. The article is useless, except for the flashes of style & tone it captures — but I suspect you & I are the only ones who can really appreciate it. The drawings were fine, although I think they fucked up on the layout — as usual — quite badly. They also cut about one-third of the article, in addition to the 4000 or so words that Don & I cut in NY. In all, a bad show, & I'm sorry it wasn't better. Maybe next time. I'd like nothing better than to work with you on one of these savage binges again, & to that end I'll tell my agent to bill us as a package — for good or ill.

Nothing binding, but certainly a notion worth
trying. The only saving grace of that Derby
scene was having you around to keep me on my
rails. What are you up to now? How did NY
pan out? What next?

In a week or so I'll send you some photos
of our main LBJ-style antagonist in the fall
election. Also my opponents for Sheriff. With
photos & some data, maybe you can rush up some
drawings for the Aspen Wallposter. In fact,
we'd use either one of those Nixon drawings
right now - if not as a cover, then as a big
inside drawing. Issue 4, now going to press,
is double-size & folded - 4 pages, in other
words; a cover, a back & 2 inside pages. We
need good art. Pat Oliphant from the Denver
Post has said he'll do a cover for us. I'll
see him this weekend in Denver, at a formation
- meeting for the Radical Journalists Union,
or some such. He said he was looking for you
in London that same weekend when you were in
Louisville with me. Strange Irony - since he
was the first artist I called to work with me
on the story. He said you were one of the few
artists in England he wanted to meet . . .

OK for now. I'll send you the fotos & other
data for the drawings I mentioned - but in
the meantime, send us anything you can't sell.
Or, for that matter, anything you feel would
be a good sort of interior advertisement for
you inre: the U.S. Press. We're constantly
sending Wallposters to editors in NY, SF, LA,
etc. So a heavy weird drawing in the Wallposter
might get you a good assignment somewhere -
maybe not. I can't say for sure. Why not get
Private Eye or The Times to send you over here
to cover my Sheriff's campaign - a Steadman-eye
view of small town politics in the American

Rockies. In fact, that sounds good enough to send to my agent. If you haven't picked up anybody to represent you, let me know & I'll see if Lynn Nesbit from IFA wants to handle your act. She's about as good as they come, I'm told. She has Tom Wolfe & that sort of thing. Even me. So let me know - on all fronts.
    Ciao, Hunter.

As it happened, I had just begun work on my second *Alice* book, having published *Alice in Wonderland* in 1967. Interestingly, the flat we were renting was full of mirrors which were having a strange effect on my mind. Reflection became a strange presence and suited my new project perfectly – *Alice through the Looking Glass*. It was vital in the production of the drawings and I was soon obsessed with the book.

Then another letter arrived from Hunter on 6 July. He was busy running his *Wallposter* broadsheet, a campaigning propaganda handout he was using in his bid to take on the 'corrupt law and order' Sheriff of Aspen in the upcoming local election. He would be standing on what he called the Freak Power Ticket. He

wanted drawings from me and said I had been listed as the
'Chief of the London Bureau'. I was flattered. He had been
impressed by my perverse use of women's lipstick and eye
make-up to colour my drawings and felt that the approach had
caught the depravity and decadence he was after in the writing.

The *Alice* project was going well. I knew I was pushing bound-
aries. It felt good. My old friend Michael Dempsey was an editor
at MacGibbon and Kee and gave me a contract to continue.

Then a letter arrived from Hunter dated 18 July 1970.

'Prepare yourself,' he wrote. 'I suspect we have struck a
very weird and maybe rich vein.' One of his friends had seen
the Derby piece and suggested: 'Hey, man!' – bound to have
said: 'Hey man!' – probably shouted as they do – some loud-
mouth in the Woody Creek Tavern, outside Aspen, in a tartan
shirt and ripped jeans, undoubtedly someone I now know as
well, and even like. 'Hey, man! – Why don't you just travel
around the country and shit on everything? The two of you
just go from New York to California and write your venomous
bullshit about everything that people respect!' Then a list of
events and places were reeled off:

```
a   series   of   Kentucky   Derby-style   articles
('with Steadman,' he added, in parenthesis)on
```

```
things like the Super Bowl, Times Square on
New Year's Eve, Mardi Gras, the Masters golf
tournament, the Americas Cup, Christmas with
the Chicago Police Commissioner, Miss America
in Atlantic City, the California State Fair,
Grand National Rodeo in Denver . . . rape
them all, quite systematically, Sell it as a
book! Amerikan Dreams — Ah, yes, I can hear
them weeping already. Where will the fuckers
show up next?' Where indeed? Ponder it, & send
word . . .
```

Warren Hinckle, at *Scanlan's*, replied damn quick. 'Yes! Let's try it!' Warren suggested that we call it *The Thompson–Steadman Report* and bill it right from the start as a long and awful series.

Then Hunter checked his optimistic generosity and snuck in sagely with: 'Or maybe "Steadman–Thompson" . . . heh . . . I can't quite recall.' I got a call from him and he wanted to get going right away. He was full of joy and I became intoxicated by this. There was something about the project which was more life-enhancing than just another job.

In a letter written in July, Hunter described the prospect as a 'rape series king-bitch dog-fucker of an idea'. He continued:

```
    We could go almost anywhere & turn out a
series of articles so weird & frightful as to
stagger  every  mind  in  journalism.  'As  we
buckled down for the approach to New Orleans
I snorted the last of our cocaine. Steadman,
far gone on acid, had locked himself in the
men's room somewhere over St Louis & the head
stewardess was frantic. I knew I would need
psychic strength & energy when we landed — to
meet the press limousine & get on with our
heinous work . . .'
    Can you grasp the lunatic possibilities of
```

42

such an assignment? Pure madness . . . on a
scale hitherto unknown . . . we could travel
with courtesans & bearers, rushing from one
scene to another in a frenzy of drugs & drink.
Indeed . . .

The only problem is that I told Warren I'd
decide on two definite scenes for us to deal
with between Aug 1 and Sept 30. It's obviously going to cost a fantastic amount of money
to bring you over here and keep you living
well for long periods of time - so our trick
is to settle two 'things' to rape in the space
of 5 or 6 weeks.
(pause 40 minutes for call to London)

Well, so much for saving money. Good to
talk to you again. Dawn is breaking here &
I guess you're looking at noon. Very strange
to jump all that way with a small black
instrument . . . Anyway, that Labor Day picnic
in Detroit idea sounds better & better,
particularly if we can focus it on 'What
happened to the American Labor Movement?'
Almost like the Derby gig, except with a
different breed of decadence. Fantastic art
possibilities, but a bit more difficult on
the writing end - needs research, old quotes,
dead dreams, etc. But as a feature I think
it looks strong. And very timely, in terms
of U.S. politics.

So the next step is yours - select a good
scene over there, either England or Ireland,
that we can handle in August, or even in late
September if necessary. We can work the travel
either way; I could come over there in Aug.
and work on the first article, then we could
come back over here for the second (& maybe
a third) . . . or else you could come over
here first for Labor Day in Detroit & maybe

one other then we could zip back to England for a third piece.

But logistically - in terms of my work schedule & my sheriff campaign & also for maintaining Warren's interest - I think we'd be far better doing the first piece over there, then getting back here by Labor Day. I'll check with Warren tomorrow & let you know how he reacts. But in the meantime it's important that you come up with some weird project that we can do over there. If you get a good idea just send it by cable to me, C/O Warren at Scanlan's in NY; that way, it'll get to both of us, quick & cheap.

OK for now. And again, it was good talking to you. Let's focus very hard & nicely on this thing - like Zen masters, or NY pawnbrokers. I can have my agent arrange the finances for both of us, if that suits you. Or we can work it out separately - or any other way. I really don't give a fuck. It looks like excellent fun, & with things going as they are, I suspect we'll be needing some of that.

Ciao . . . Hunter.

That we got some way down that road is testament to the soundness of our judgement and our terrible ambition, but that we did not go further is more to do with the vagaries and fate of magazine publishing coupled with the ambitions of others and, perhaps, to a lesser extent, the uncertain commitment of our collective drive. Our respective lives were so different. Nonetheless, back then we were willing enough to go just about anywhere. There was a point when I suggested *Fear and Loathing on a Slow Boat to China* and Hunter replied: 'Goddammit, Ralph! I knew if I waited long enough you would come up with something truly monumental' – but we didn't get around to that one – or *The Highland Games* or *A Visit to the Queen of England*, which, strangely, some years later, became a

1970. Scenes of the Kentucky Derby, including showing Ralph in toilet mirror, Warren Hinckle, with eye patch, and Sidney Zion (editors of *Scanlan's*), the press room at the Derby and spectators.

Drawing of search for the artist to be used to illustrate *Fear and Loathing in Las Vegas*.

Original spread for first instalment of *Fear and Loathing in Las Vegas* which appeared in *Rolling Stone* magazine in November 1971.

Scene from *Fear and Loathing in Las Vegas* of the police chiefs at a drugs convention.

1977. Hunter and Ralph in the desert between scenes for the film *Fear and Loathing on the Road to Hollywood*.

Dr Jo Sterling Baxter, Republican candidate for Pitkin County Commissioner and Hunter opponent during his sheriff 'freak power' campaign.

TAKE
ANOTHER
LOOK AT
YOUR
SHERIFF —
AND **VOTE**
BEFORE HE
SHOOTS

Ralph's drawing for Hunter's campaign for sheriff.

The rest of Hunter's opposition.
A. incumbent sheriff Carrol Whitmore, his Democratic opponent.
B. Glen Ricks, Republican candidate opponent.

1972. Scenes of fishing picnic at the Roaring Fork River, Woody Creek, Aspen, Colorado, with Ralph, Anna and Juan Thompson, Hunter's son.

1972. Intimate studies of Hunter and
Sandy at Owl Farm, Woody Creek.

Patriotic cabin on road to
Woody Creek.

Front entrance of Owl Farm.

2005. Memorial flag at Owl Farm
at sunset.

1973. Hunter poolside
at the Hilton Hotel,
Washington, during the
Watergate hearings.

1981. Hunter at the Sheraton
Hotel, Washington.

1973. Hunter's first meeting
with Mo Dean, Beverly Hills.

1973. Art Garfunkel and
Jann Wenner, San Francisco.

1976. Bill Cardoso, the man
who coined the word 'gonzo',
in Sausalito, California.

1973. Anna in the regalia of a Navaho squaw during the Santa Fe Indian Festival, and Ralph in a Navaho hat.

Jann Wenner and John Dean in the hall at the Republican Convention, Kansas City, 1976.

John Dean and Ralph in Washington.

project that Simon Kelner, editor of *The Independent*, attempted to initiate as a project Hunter might like to undertake.

For a month or so there was silence again and I returned to *Alice*, which had become a major theme for me. I pursued the idea with an intensity that does not enter the spirit every day. Meanwhile, my job at *The Times* was foundering on the sharp, shoreline rocks of readers' complaints.

I got another call. Hunter had been asked to get me to meet him in New York and then the plan was for both of us to take off for Newport, Rhode Island, where, every year, muscle-bound American deck-hands struggle to keep a boat upright, competing with a similar craft and crew. Together, they zigzag a given course and hope to outwit each other, using the wind, tactics and brute strength. Each boat is worth the price of a new university and they are watched by gin-soaked yachting types, male and female, in captain's hats, lounging in deckchairs inside Perspex-covered enclosures at the front end of yet more expensive, floating country houses representing nothing more than elegantly vulgar expressions of dodgy wealth. The America's Cup.

We took a bus to the airport. By now, I don't think there was much money left of the three quarters of a million dollars with which Warren Hinckle III and Sidney Zion had been entrusted just nine months earlier. First-class travel between San Francisco and New York every forty-eight hours in the interests of journalistic perspicacity soon eats away at the edges of a small fortune.

We went downstairs to the Irish pub beneath the *Scanlan's* office to discuss the America's Cup project. Donald Goddard was with us, smoking his hand-rolled Havana cigars as usual. It was in dear Donald that I put my trust. As long as he was there at the helm, I figured, this would be a kosher journalistic venture. J. C. Suares was there, too, and he wanted some weird 'pitchers ta put in da magazine. Sumtink off da wall. Ya know what miracles are, right? We need sum o' dem! Know what ah mean? Sum o' dos tings we seen in yor book Don brawt in heah. Da sky's da limit 'cos dat's why you are heah! Udderwise, dere's no reason fer any o' dis.' His Brooklyn accent spelled out our brief.

It spoke panoramas to me of what was expected and I wondered if this might be the beginning of something unprecedented. He did not know that at the time, but his was a signal to break all the rules and go for a massive effort to take everything on board and give it the shot I had never been able to give to anything in England. In England, in spite of the so-called 'satire boom', everything was still, in its quaint way, under wraps. But now I was in a foreign land and I felt it my foreign duty to deliver something they had never seen before.

On the bus to the airport, Hunter had taken one, or maybe two, little yellow pills, not unlike diazepam. I let it pass and even assumed that he maybe had a medical condition; the kind of condition one does not ask about of someone one hardly knows. I thought he might have been grieving. I had no intimate knowledge of any American thus far and maybe all Americans take these pills to withstand the pressure of the responsibility thrust upon them as defenders of the known Free World. Perhaps I would need to take care of him. I would be prepared, I reckoned, being as I was an ex-Boy Scout and indeed a Troop Leader with proficiency badges of high degree.

We arrived at La Guardia to board a plane to Boston, where we would pick up a small internal flight to get us to Newport, Rhode Island, the setting for the race. Before any flight, long or short, I get nervous and a drink or two usually reaches the tiny innermost parts of my being, giving me courage of a kind I cannot get from religion. It became obvious that Hunter also saw fit to partake and so it was with light hearts that we boarded Flight A157 to Rhode Island.

Hunter explained on the plane that we were not staying in a hotel as was previously suggested, but becoming, instead, part of the crew on a three-masted sloop whose record at sea was not something we needed to discuss. 'Okay by me,' I mumbled, at this point more concerned by the fact that we were going to be living on a boat at all for at least a week.

'D'you like boats?' he asked me. I shrugged and laughed off my disquiet.

'They're okay,' I replied, 'but only when they're in shallow water, so I can wade ashore whenever I feel like it.'

'Well, never mind, Ralph. This is only journalism and the boats are rich men's playthings. If we fuck up, that's just part of the story, right? There will be a rock band on board for entertainment too. They have no name as far as I know and neither does the boat, for security reasons. Heh! Heh!' I nodded and looked through the small window as we approached the runway. I showed no emotion. When you are on assignment you don't ask questions. When you are building a career, who needs to know?

We emerged from a taxi right on Sayer's Wharf, the heart of Newport's jetty district, where boats bobbed and jostled to the lap-lap of waters blown by a gentle coastal inlet breeze, where all the real boats come in, or so it seemed. Back in 1970, things were still good, rough and basic, decking boards defining the fisherman's domain. It reminded me of Whitstable, an old fishing port in England, where fresh fish is the definition of a rough and healthy way of life, and still is, as long as you accept the tourists as part of the scenery. We all are. After all, Newport is probably as old as Whitstable. In fact it had a Golden Age, prior to 1776, and has the oldest library in America and the oldest synagogue. It groaned with restored colonial buildings and swelled with privileged pride. I noted this more than anything else. Beware of privilege. It stinks of rotten fish-heads, many of which were lapping the shore beneath the jetties. There was another terrace of jetties full of yachts and seamen of an entirely different type. Weekend sea-slobs who were fast becoming the elevated new cream of Newport society. They were the ones we were there to observe and report back on. Gonzo was about to rear its undefined head.

The bar on the sea-front had a piano played by someone who had probably seen more bar fights than a whiskey optic and a few more drinks there gave me enough Dutch courage to face the boat Hunter had located more or less out front. It was a vision of well-varnished old wood and worn deck-boards. Safe as houses, as it transpired. It was about fifty feet long and knifed the water in an extremely elegant way. We were welcomed aboard by a willowy bloke, our captain. He had the suggestion of a beard and wore a polo-neck sweater but he

47

was no hoary sea dog. I went below and, sitting on my own bunk, I relaxed but resolutely kept all my stuff in my case. In fact it stayed there for the entire week. There was no privacy down here and Hunter had already lain down on his bunk and gone to sleep. It was a surprise to me. There were no social niceties, no polite tittle-tattle. Just a gentle snore. He was out of it. I decided to go up on deck and look around, trying to imagine how it might feel to be a seasoned sailor.

I walked round to the prow of the boat, rolled myself a cigarette, and lit up. I puffed out my chest, held my arms behind my back and stood, legs akimbo. I looked out to sea beyond the boats in the harbour, gazing upon the flotilla of ocean-going luxury yachts moored out in the bay, drooling money. I could just make out the social scene inside the Perspex-covered lounges, men in captain's hats, navy blue jackets, golden ephemera, city men tarted up for the occasion, sporting big cigars and clutching glasses of gin. Lounging women of all shapes and sizes, whorish, glamorous, bosomy, good-time, as well as others who had managed to hang on down the years with face-lifts and plastic surgery. It was a predictable scene and part of the reason we were there. They, as well as the crass, Olympian sea gods who manned the two boats taking part in the annual event, the American *Intrepid* and the *Gretel II* from Australia, were the enemy. I immediately felt like an outsider, which was good, of course. I needed the buzz that came from a 'them' and 'us' confrontation in order to inject any drawings I might produce with a strong sense of corrosive contempt.

While I was honing my rebellious inclinations to fever pitch, a new group had just come aboard. Four guys with fashionably long hair heaved their luggage and what looked like an orchestra of instruments aboard. I turned and through a haze of 'Hey, man! How ya doin'?' and clasping of bravado-driven handshakes between men on a mission, I met the rock'n'roll band. I remember thinking that they were about as familiar with back ends of boats as they were with the front. This was the rest of our team – the distraction, and if that wasn't the name of their band it damn well ought to have been.

They were shown down below and galumphed their way

down into the crowded berth. This reignited my disquiet but I let it pass. Hunter was awake now and beers were passed around as we got to know each other. Whether we got pissed at that point, I don't remember, but within the next eight hours we certainly did.

We remained incarcerated in this hell-hole for a week, during which time we attempted to entertain the enemy in between yachting jousts with loud music and me — yes, me! — on bongo drums. We had transformed ourselves into a ship of dangerous fools who turned out to be the only entertainment in the bay.

Out at sea, in the area known as Rhode Island Sound, where the races were taking place, the wind had dropped to almost zero and the yachting jousts were slow. Almost nothing was happening, particularly for the gin set who had been lured onto their boats by enthusiastic diehards with promises of limitless drink, which was true, and exciting demonstrations of fine yachtsmanship by monarchs of the sport, which was not, unless, of course, you were in the race. For a spectator, it was all so boring. When there is no wind, nothing moves and nothing much happens, except for a lot of fumbling around on-deck by sunburned rednecks trying to get the wretched boats to move.

Motorized, we cruised in and out of these daft buggers, the rock band having a whale of a time. The gin crowd loved it and cheered our pirate ship whenever we appeared. At nights, moored now centre-stage in the harbour, we — and I mean we — gave rock concerts every night to a coastline alive with night-life. Bongos became my art form for the week and marijuana became a weed of great expression for me for the first time in my life. I became hooked on the cheers that erupted from the shore. We were the real champions. Energized fans clambered aboard from small dinghies, wanting to become part of the action. The women hung on my bongos and wanted to learn how to become a star. They couldn't, of course, but they wanted to be part of the action — and the marijuana.

But it didn't last. The band wanted to pack up and leave, sensing a sinking ship. Rock bands are the first to know when the gig is over, man! — and Hunter and I had no story. Worse than that, I had been suffering from a queasy stomach for the last couple of days. Every time we sailed out to see if the race had started, I began to get seasick. To try to keep it down I took the wheel and stood upright against what was becoming a dull and rainy seascape. I recited to myself: 'The boy stood on the burning deck, whence all but he had fled, my stomach tells me all is lost, and I should be in bed.'

For a lot of the time Hunter remained below deck and slept. The roll of the boat had no apparent effect on him although he was still gobbling these little yellow pills like sweets.

It happened on the Friday, our last day during the finals or whatever they were. There had been no fierce action from the racing yachts and some of the luxury boats had started to leave. Hunter appeared and suggested a walk along the jetties and mooring bays as a change of scene and maybe a talk about what our strategy would be. Since we still had no story, an idle walk on dry land, and an inspection of the boats, especially the two racing yachts and where they were situated at night, might give us some hint of a wild plan. We were also running out of money and a call to the *Scanlan's* office might be a good idea. As it turned out, it was the worst idea as we were calmly told that the magazine was folding and that this was to be their last issue, ours the last story — if we had one, that is!

We walked idly in and out of quayside shops, inspecting all manner of yachting goods, clothing and equipment. It was in one of these shops that Hunter picked up a jack-knife with one of those rounded 'pig-sticker' blades, used the world over by Boy Scouts for extracting stones from horses' hooves. I was extremely familiar with them. Hunter bought it there and then.

'Whadya buy that for?' I asked.

'Just in case!' he snapped

'Just in case of what?' I persisted.

'Who knows, Ralph. I bought it on impulse.' I didn't pursue it any further. We did our tour of the jetties, and noted where the racing yachts would be moored when they came back from the race. Hunter had taken careful note, at the same time asking a few idle questions about how they bring them out each morning, and would there be a final public display tomorrow before they were wrapped up for another year?

I learned that each boat was twelve metres long, and America's *Intrepid* was not only the favourite; it was revolutionary. It was the first 'Twelve', as they referred to them, to separate the rudder from the keel, build in what they called a bustle or kicker, finishing the keel off with a trim tab. The designer, Olin Stephens, who was well-respected world-wide, had built this elegant craft out of double-planked mahogany on white oak frames and no expense had been spared. It was state of the art – yessir! It had won the race in 1967 and was going to kick the butt of the Australian challenger, *Gretel II*, that very day. 'You ought to be out there right now and watch history in the making,' they said, but I was all watched-out and still sick. Hunter, on the other hand, was thoughtful.

We strolled back towards the main cluster of shops and bars. 'Fancy a drink?' I said. 'I'm still feeling sick but sometimes a shot and a beer just hits the spot.'

'Okay. Why not?'

'By the way,' I said, 'what are those little pills you keep taking?'

'They keep me sane and hopeful. Why?'

'Well, you never seem to feel seasick, but I thought you had some kind of condition.'

'Fuck, no! I feel as right as rain.'

'Well, maybe they would cure my seasickness,' I said.

'Maybe,' replied Hunter, 'but you only need one.'

'Okay,' I said. 'Let me try one.'

'Up to you,' he said, handing me one. I swallowed it and asked: 'What happens now?'

'Nothing,' said Hunter, 'for about an hour. Then you may feel a little weird.'

'Okay,' I said, 'let's get a drink.'

The bar was filling with the Friday evening crowd and the pianist was playing requests. 'You're the Cream in My Coffee', 'Let It Be' and 'Listen, Do You Want to Know a Secret?' were just a clutch of the favourites being played. I remember a couple of dogs lying under the piano looking bored. Obviously yachting dogs, lashed to the boom, squared and tied to chair legs like accessories for rent. Splice the main brace, Jim lad! Give the dog another ship's biscuit! That's more like it. Drinks all round! Toddle the pissle! What? Cringe the pound. Griddle the Poooooooo-o-o-o-o—o—o-rk-oooohho-o—o—o-arck − wait − arcgrhhhgrh − what?? HooontrrassTomgbutt. My fairy issssss—look over thereee − aaarrrghhh. Pull yusself togertherrrrrdddd − Wha?!!! There are snarling, red-eyed dogs eating my socks . . .

I had begun to feel weird. Richard Nixon's brother − at least, I think it was his brother − was leaning diagonally against the piano, talking nonchalantly to other red-eyed yachting types and paying no heed to the red-eyed, snarling dogs I was trying to avoid. People were beginning to melt and I felt incredibly good. Wherever I walked, specific salty types, tough as they came, melted out of vision. At that moment, I was a king. Hunter was there, but he was no longer significant. He was watching me, and he was following me like a nanny. Next thing I knew, we were outside looking at a perfect moon hanging over the harbour.

'I think it's time we got back to our boat. We have things to talk about,' Hunter was saying

'Okay-arrrrrrrrrarghhhhhhh!' I replied. Weeeeeeeee-sssssstilll neeeeed a sssssssstoooooooooryopp! Hah!!'

'Settle down, Ralph. We have a long night ahead,' he said, eyeing me curiously.

Hunter was good at manipulating oars in a clumsy kind of way and it seemed natural as he rowed us back to the boat. Soft water lapped at the bulkhead and all was quiet. In fact, the entire bay seemed quiet. The intense moon pierced the sea with shards of light that were beginning to play strange tricks on my eyes and my senses seemed unnaturally alert. I was wired into every detail of every surface I looked at. I saw nailheads in the bulkhead which were gleaming from the moon's light. Metal boat fixings shone like beacons and rigging glowed, looking like electrified wires that were on fire. Every touch on any surface felt like the skin of a dinosaur. My senses had been warped through ninety degrees. As I looked along the deck, I thought what I was seeing was a plunging cliff face. Hunter steadied me and seemed to be saying: 'Keeeeeeep goooing, Raaaaaaaaaaalph! The cabin's just here.' I grabbed an upright, the door-frame, and groped my way inside but stayed on the floor.

'Okay, Ralph,' said Hunter. 'You're having a rush. It'll pass. Listen to me! You are having a rush. Waaaaaiiiittttttttt-Hooo-steeeeaaady. Iiiit wiiiiilllll paaaaaaaasssss.'

I don't remember how long this feeling remained but gradually I began to realize that I was on some terrible high that was not drunkenness. I knew that my mind had been hijacked and I was on this hideous beast for the ride.

'Wharwe gonna dooooo?' I could hardly speak and tried to lift myself up to sit on something other than the floor. Hunter was there, but also wasn't there, supporting me into an upright position, and he didn't, but did.

'I have no idea whaaaart we are goooing to doooooooooo, buttttttweeeee nnnnneeeeeeeeeed a plan-oooooorrrrrr weee-eeeeee're fuuuuuuuucked right now,' he seemed to slur.

'I neeeeedddd sooome aiiir riiiight now,' I said, gripping the door frame and falling outside into the air. I put a finger of my right hand up in front of my eyes and watched it, very specifically, turn from a finger into the Statue of Liberty, a ship's bollard and then a finger again. It was at that moment that I knew we were into something I had never known before. This wasn't just pissed; this was another world. I was strange. I put my hand up to my forehead and wiped my brow. 'Jesus!' My hair – I still had some then – felt like Hitler's! I was Hitler in that moment and as evil as anything he could ever have wished to be. I was trapped inside some heinous plot that grafted itself onto my captured paranoia like a limpet mine. At that moment I knew I was capable of anything. I was calm. I knew I had something important to do for whatever cause I was here to support. I fumbled my way back inside the cabin. 'Where's the rock group?' I said.

'Long gone,' replied Hunter. 'Thank God you didn't leave your passport and your ticket home behind in your suitcase.'

I fell down into the hold and scrabbled around inside my suitcase. Most things were gone and instinctively I felt around my person. I had my ticket home and my passport, secreted in an inside pocket with a few dollars. Most of my other gear was gone, stolen by the band, and the only other thing they had not taken were the socks on my feet. 'Where's Andy?' I asked.

'On shore, somewhere. But that is not the point, Ralph. We still have a job to do.'

'Oh, the job,' I said. 'That's easy! Let's do it now. I'm up for anything.'

'Anything?' said Hunter

'Anything,' I said, enthusiastically. 'You bet!'

'Okay, Ralph. Well . . .' He paused. '. . . you remember walking around the jetties this afternoon, seeing where the boats are moored and all that?'

'Yep! Remember that. So what we gonna do about that?'

'*You!* Ralph! What are *you* gonna do about that?'

'Me!' What can I do? I'm only an artist!'

'Yes! Ralph! You are only an artist, but you are full of ideas.' He stared hard at me.

'Okay. So . . . ?' I was confused.

It was at this moment that Hunter produced two spray-cans – one black and one red. 'I bought these,' he said, 'while I was waiting for you in New York. I wasn't sure if we'd need them, but right now we have absolutely no story.'

'Shit! You're right. I had forgotten about the story. So why the cans of spray paint?'

'Well, you're the artist Ralph. What do you suggest?'

'Something crude on a wall!' I suggested.

'Okay,' said Hunter, 'but what? You're the artist from England. Something about the IRA, perhaps?'

'Nah!' I said, 'How about if I spray-painted "FUCK THE POPE" on the side of one of those fancy yachts. It is almost unthinkable. No one could imagine such a thing.'

'Hot damn, Ralph! Are you a Catholic?'

'Nope! Why? It was just the first thing that came into my head.'

'Okay – just wondered. But if you're game, let's do it. I reckon I can row that little dinghy in between the jetties and get us between the two racing yachts, then the rest is up to you. In the morning when the crew sail out into the harbour, only then will they see your scurvy act of vandalism and blasphemy.' A gentle smile crossed his face and his eyes gazed into the middle distance.

'All right,' I said. 'Let's do it!' By this time trapped inside the drug's reverie I could have sprayed out Michelangelo's *Last Judgement* on the ceiling of the Sistine Chapel. A yacht would be a beggar's handcart by comparison.

We now had to clamber into the moored dinghy and Hunter would do the rowing and sculling. I held the spray-cans. My mind had become a clear mechanism of wilfulness. Hunter proved to be quite a boatman, skilfully manoeuvring our little craft in and out of the jetties. The reason for our afternoon constitutional was becoming clear and logical to me. I instinctively knew that what we were doing was not good but my drug-induced state overruled anything but this specific mission. The Guns of Navarone had to be destroyed at any cost, do or die, for God and country. It was them or us.

We had arrived at the two jetties between which, silently moored and battened down, were the two racing yachts, the *Gretel II* and the *Intrepid*. A chain was strung between the jetties, a 'Keep Out' sign hanging from it. Hunter was now very quiet and signalled to me, with a finger to his lips, to be likewise. The moon was a pure white ball but the flashing reflections in the gently undulating sea were strangely red to me. There was red on one of the boats and I must have been seeing a duality of these elements. One of the boats was slightly in shadow and Hunter quietly steered the dinghy towards it. We wobbled between these two amazing machines which now resembled floating jack-knives like the one Hunter had bought that very afternoon.

'Okay,' he whispered to me, 'it's now or never.' I held the black spray-can up in front of my face and shook the can, as one does, to mix the contents. There was a terrible, loud clicking sound from inside the canister – the last thing we wanted to hear.

A voice from above on the jetty barked, 'Hey! Who's down there?'

Hunter cursed and snarled in a tight whisper: 'Fuck! No, Ralph! We are doomed! We've failed. There will be pigs all over the place in a minute. We must flee!' A torch from above picked us out and the voice shouted, 'What are you doing down there? This is a cordoned off area, didn't you see the chain?'

'Oh, hell no! We were just looking at the boats, just looking at the boats,' Hunter called back. Then he whispered to me again. 'We must flee, Ralph. There'll be pigs with guns and dogs crawling all over this place. We must flee.' He desperately pulled on the oars, trying to move backwards and out of the area. One of the oars came out of its rollock and I have this abiding image of Hunter's shorts and his bare legs stuck up in the air with his back on the deck. He flailed, trying to right himself, cursing and muttering all the time. He pulled again on the oars and we slipped out under the chain and out to sea into the darkness amongst the other moored boats.

'Pigs everywhere' and 'We must flee like hunted animals,' he kept saying, and then he decided: 'We'll hole up on the sloop until we can make it to dry land.' No one was following us yet and, in all honesty, nobody probably did. They could not imagine what foul deed we had been about to commit. They must have thought we had merely been looking at the boats, as Hunter had said, and I had dropped the foul evidence of our intention into the water and we clambered aboard our vessel.

Hunter had taken out his wirebound notebook and had started to write things down. Meanwhile, I had begun to tremble uncontrollably as flashes of sense returned to my addled brain. I started to gabble like an idiot and Hunter wrote it down feverishly. 'That's good, Ralph. What else did you see? Great! What were you thinking? That's good. Go on. What else?'

I gabbled on automatically, shivering with shock and horror at what I had so nearly done. Hunter continued to write, urging me to continue. I was clutching my shoulders tightly and lay curled up on the cabin floor. Now, getting out of here was uppermost in my mind. I was coming down, but I was also scared.

'Get your case, Ralph! We'll hail a passing boat coming in − but first we must make clear our failure!'

'Whadya mean?' I said. From out of his general junk bag he took a pistol with a wide barrel.

'What the hell's that for?' I asked.

'It's a Leary distress flare gun,' he replied. 'It will light up the harbour and people will be so startled to see one of these buggers there that they will be too freaked out to notice us, and we'll be outta here!' Red and green flares left the gun in quick succession. But one of the still-flaring distress signals reached the top of its arc and fell unerringly onto the wooden deck of a yacht that had until that moment been in no distress. Bedlam erupted around the bay. Fires on boats are serious matters, particularly in a safe harbour, but more so when an SOS signal had only moments before alerted the town of Newport that something was up. Those things are capable of lighting up square miles of an area in mid-Atlantic and are a matter of life or death. They weren't the hand-held waving variety, as fire safety would be the last requirement with miles of water all around. In the general panic − a fire engine had already arrived on the scene − Hunter hailed a late-returning fishing boat like a desperate survivor. That there was one passing right then was a sheer fluke. Thankfully, no one was asking any questions at that moment. It was all hands on deck and we looked as though we were in trouble.

I grabbed my bag and the fishing boat hauled us aboard and took us into a jetty. It was now dawn and we holed up in an early-morning coffee shop where people were talking about 'some fucking maniac' setting fire to boats. I was feeling like a whipped puppy but was doing my damnedest to be philosophical. Hunter was immediately on the phone, looking busy, calling Aspen and registering for the Aspen Sheriff's election. He also hired us a Cessna light aircraft on his Amex card to get us out of Newport and back to Boston.

Meanwhile I was still descending from my God-awful experience. I had lost my shoes, and most of the stuff in my case. The plane ride to Boston was still within the bounds of reason to my confused mind and I nodded off for a while, but awoke startled whenever the plane hit an air pocket. But paranoia strikes deep in the heart and when we landed I immediately began to look around for signs of police patrols looking for two crazy characters who had just tried to write 'FUCK THE POPE' on the hull of a multimillion-dollar yacht. I lurked behind pillars and hid in the toilets. Hunter was on the phone when I found him again, ringing the New York office, trying to get someone to meet me at the airport when I arrived. He was describing me as a 'basket case'.

'If someone doesn't meet him he is likely to get himself arrested as an anti-social foreign vagrant. He has been raving incoherently for ninety straight hours and I think he is likely to take on the entire Boston police force. He vows he will never set foot in this country again if he doesn't get justice. He still thinks he's in the colonies. Someone has to meet him and calm him down. I have to get back to Colorado and get involved in civic duties.'

It was Saturday and the office was closing at midday. *Ninety* hours? I had been raving for ninety hours? Hunter assured me that I had. He made his excuses, saying that he had to go but he reassured me that someone was going to meet me at the airport. We said our goodbyes and he assured me yet again that there would be someone to meet me. Then he went off through a gate to catch his plane, leaving me to wait another couple of hours before I could get a flight.

I felt like a fugitive on the run and either sat at a bar with my head in a beer or shuffled past police security behind pillars or mingled with groups of people moving purposefully towards flight gates going to Atlanta or San Francisco. My eyes were wide with terror and lack of sleep. I felt like a guilty fugitive, a miscreant who had no right to be on the loose. 'I haven't done anything,' I kept telling myself, but the sight of anyone resembling an official sent me into a cold sweat, a blurred panic and a reflex action made me clutch at my passport and ticket home which lay, thankfully, in an inside pocket. My flight came up on the departure board and I seized the moment to begin the walk to the gate. I slowed down, trying to control my schizophrenic behaviour. Then I stopped dead and realized that I hadn't checked in my suitcase. It was light — suspiciously so because, of course, there was hardly anything in it! Dirty underwear and a sketchbook. No socks! Not even on my feet. No shoes and the police were looking for me. Had to be. I was on the run. Get a grip! Just walk to the check-in desk and try to be charming, using my most English accent to appear respectable. If they thought I was eccentric, so what? Lots of English visitors walk about with no shoes.

'Is this your luggage, sir?'

'Er, yes — such as it is. Had a spot of trouble on a yacht — washed overboard, most of it. Damned new suit went over too.'

'First time?'

'First time, what?'

'On a yacht?'

'Oh, that. Huh! Never again. Not cut out for it I guess. Can't wait to get home.'

'I'll bet. Have a nice flight.'

When I got on board there were two things I could not do. The first was sit down. I insisted on standing at the back, near the galley, clutching the back of a seat. Today, my behaviour would have led to me being escorted off the flight and arrested, but this was 1970 and Laila Khaled had only just carried out one of the first hijackings when she took control of a plane in the Middle East and redirected it to Damascus. In spite of this, it was still possible to display strange behaviour in an aeroplane without connections being made and terrorist activities were still a remote possibility. Neither were there many in-flight safety regulations against insisting on standing for the entire journey. I pleaded a rare condition that meant I could never sit down when on the move. My bloodshot eyes and terror-stricken expression reassured the air-hostess that I was sincere. And one other thing – I couldn't close my eyes without seeing moving purple flesh pulsating under my eyelids as well as veins of some green substance which was puzzling medical science – and my doctor had advised me not to fly. Thank God I was on my way back to England, I said. The air-hostess smiled nervously and showed sympathetic concern, saying that if there was anything I needed I should just press the steward's button above the seats.

It was a short flight and I held on to that seat all the way. I was really no trouble. My bag entered the hall at the same time I did and I rushed out to see if J. C. Suares or anyone from *Scanlan's* was there to meet me. Naturally, no one had turned up.

I went to the taxi rank and hailed a cab. I asked how much it would cost to take me into New York and luck was with me. It was twenty-three dollars and I had twenty-five, which had to be a gift from God. When I arrived at the *Scanlan's* office it was closed, of course, and I went downstairs to the Irish Bar which, fortunately, was open. When I went inside, the barman whom I had befriended before was on duty. He commented on my wild appearance.

'Yeah! I know,' I said. 'I am in a bit of a fix, man. Could you lend me a quarter to make a local call?' He did, bless him, and I phoned the only number I knew in New York, which was the home of a lovely lady called Ann Beneduce. I had met her in Bologna at the Children's Book Fair, the previous year. She remembered me because I had put her in hospital by over-turning a car in a ditch on our way back to her hotel after a party. She had broken a couple of ribs but was okay now. She was just on her way out. I gabbled my predicament, which must have transmitted itself down the line because she simply said: 'Just get in a cab. I'll pay this end.'

The barman called me a cab and I climbed into the back,

by this time palpitating and throbbing like an open wound. When I arrived Ann was standing at the door of her apartment. 'My God! she said. 'You're blue! I'll call a doctor,' which she did immediately. I fell onto a sofa and the next thing I remember was a doctor sitting by my side with a syringe in his hand.

'I am going to give you a shot of Librium,' he said. 'You have a serious case of hyperventilation. You are very lucky. Left much longer, your heart would have exploded! Are you an . . . er . . . regular kind of guy . . . y'know, do you do this often?'

'Hell! No! Of course not! I have four lovely children. I am devoted to them, although I am going through a painful domestic situation.'

'Ah!' said the doctor, 'that could explain a lot.'

'Explain what?' I erupted. 'Just because I did some drug I had never heard of before and have ended up like a gherkin on a sofa in New York. Of course I am a regular kind of guy, but I guess I do have proclivities that I keep well-hidden. Hallucinogens mean nothing to me. Drugs of this type, including lysergic acid diethylamide − LSD, I guess − containing indoles found in mushrooms and the innocent Morning Glory seeds, can be bought in any wholesome gardening emporium. In the wild, they proliferate as nature's own free life enhancers. Psilocybe cubensis, mescal or peyote are there for our use. Unfortunately, mean, greedy poncers from all walks of life, including politics, distil these natural items into dangerous alkaloids to sell for personal gain. Did you know that the innocent coca leaf is an invigorating stimulant which can slow the metabolic rate of the sickest person and give the system back its fighting balance to restore itself and ward off fever symptoms and if necessary alleviate altitude sickness, but distilled into the alkaloid, cocaine is the perfect example of an innocence transformed into a . . . into aaar . . . socia-l de-es-e-a-s-ezzz. In fact . . .' ZUMPF! I went out like a light.

I must have been out for twenty-four hours because it was Sunday afternoon when I came to. Ann had saved my life and I shall be forever in her debt. I was calmer now and had some light food. My explanations must have sounded odd to her but this was not the first time we had been thrown together so dramatically. I related my story as best I could and it didn't surprise her in the least. In fact I made a drawing that very day with the racing yacht looking like a jack knife in the water in the moonlight, the FUCK THE POPE graffiti gleaming against the sky and the aura of vicious red ink permeating everything.

This trip, possibly more than any other, established a pattern of journalism, if that is what it was, that cemented my friendship with Hunter and laid the ground-plan for future assignments which, by their very nature, could only be classed as pure Gonzo.

When I returned to England a couple of days later, I went into *The Times* to make a drawing and fell asleep across my drawing board. Simple jetlag.

And now, thirty-five years later, I remember the America's Cup well, the particular America's Cup, the America's Cup of my nightmares and daymares, ugly flashbacks and palpitations, dream-scarred eyes behind pulsing flesh lids, trying to sleep, trying to forget the aftermath of what appeared to start as an intriguing and pleasant week.

Dr Hunter S. Thompson – ah yes! Him and life on the ocean, a time-honoured code of seafaring ways, no cares, free, at sea and awash with good buddies, for a romp on a fifty-foot sloop, seasick but sod it. It was a defining assignment of Gonzo and work was the last thing on our minds. *Scanlan's* magazine was going down anyway, but the sloop was afloat and we were on it with enough of their money to keep us flying, at least until the end of the race, that is if they didn't keep stopping to review the positions of the two racing yachts fighting it out for some advantage on a heaving Atlantic swell. There is no knowing what you do with a nervous captain in charge of two freak journalists and an unknown rock band along for the ego trip, except maybe ignore him and hope that he understands the irrational behaviour patterns of artists at work. I don't remember how he did take the strange downhill turn when I finally overcame my miserable three-day seasickness with a pill that drove me from reasonable consciousness to wild and dribbling vandalism, intent on only one act, an act so unthinkable as to render it impossible under any other circumstance. It remains a defining moment in the evolution of Gonzo and, without doubt, a dress rehearsal for *Fear and Loathing in Las Vegas*. For Hunter, it provided living proof that going crazy as a journalistic style was possible. Those *Fear and Loathing* drawings were only possible for me because of the America's Cup six months earlier, which injected the drawings with the eerie sense of being there to record the sensations. It was a regurgitation, a psycho-artistic vomit – a creative, cathartic cleansing of my inner being.

The next six months were a combination of social skills and trying to look normal. I lost my job at *The Times* for what was referred to as mild sedition – a contradiction in terms, I would have thought. Perhaps they meant the mild indifference of the reader response to the seditious nature of my work. I had always combined untroubled law-abiding nerdiness with a decent community profile and was an ambitious young man with a mind-bending desire to succeed at any cost. So, it was no problem at all. If it wasn't for the mind-flashes and troubling surfing of nocturnal imagery, I could have withstood the psychological turbulence and eventually got through it. Unfortunately, the turbulence did not subside and I drank a bit more than usual. Anna was my constant support and seemed to understand my predicament. She knew that something dramatic had happened but, to her great credit, she never pushed her curiosity. I hadn't heard from Hunter in a while, found editorial work where I could, continued the work on what I have to say became my classic version of *Alice through the Looking Glass* and happened upon a bookshop in Kensington Church Walk, London W8, run by a wonderful little man called Bernard Stone, who became a lifelong friend.

It became a regular haunt, a second home, a watering hole for poets and others and ultimately for me an outlet for printing, on an old flatbed proofing press, very limited editions of illustrated broadsheets by dead and living poets ranging from Stephen Spender to Sylvia Plath. I would work on *Alice* in the morning in our new home on the New King's Road and then drive over to see Bernard in the afternoon and we would plan another broadsheet. He was always being presented with poems by just about anybody who was anybody, including Ted Hughes, Lawrence Durrell, Christopher Logue, Brian Patten, Adrian Henri and even some who were not really poets at all.

Hunter was one of the latter kind, though I am sure he made it not scan on purpose: I slip one in here even though he sent it to me years later . . .

·HUNTER S.THOMPSON

September 30, 1994

To Ralph

I TOLD HIM IT WAS WRONG

By F.X. Leach

Omaha 1968

A filthy young pig
got tired of his gig
and begged for a transfer
to Texas.
Police ran him down .
on the Outskirts of town
and ripped off his Nuts
With a coathanger. . . . . . .
Everything after that was like
coming home in a cage on the
back of a train from
New Orleans on a Saturday
night
with no money and cancer and
a dead girlfreend.
In the end it was no use.
He died on his knees in a barn
yard
with all the others watching.
Res Ipsa Loquitar

But it was fun to make a drawing for whatever the poem might have meant, have a simple line-block made and to set the type myself. Then we could choose all kinds of hand-made papers and jacket fabrics and arrange a signing of about fifty copies. Bernard would sell them to collectors for a song, mainly to America and I don't think we ever made more money than we were spending on materials.

At about six every evening, Anna would arrive at the shop and we would consume dreadful red wine at 'Chez Bernard' for an hour, close up the shop and repair to the Elephant and Castle pub around the corner, where we discussed where we were going to eat that night. At three in the morning we might still be in the restaurant, singing songs or boozing again at a nightspot called the Meridiana. It was a carefree period and we took to it like genuine lowlife. I had also found an agent to handle my book illustration work and bought a ruin of a lean-to in the Languedoc region of southern France for five hundred pounds, which Anna and I spent the next fifteen years making habitable. However, I was still getting flashbacks of my America's Cup experience, which remained unexpressed and unresolved, except for the couple of drawings I made in New York after the Librium sleep.

FEAR AND LOATHING
Summer 1971

*Drawing with the right kind of eyes ... A Chicano
lawyer takes my place ... At last I draw a motorbike
... Rolling Stone flips, freaks & goes crazy*

Early in 1971, I got a letter from Hunter, who had finally
tracked me down to our new home. With the letter was a
gruesome manuscript he had been working on – 'Something
experiemental,' he said, and would I be able to do some sick
drawings to express the awfulness of what he had been
through with a man called Oscar Acosta, a 'Samoan Lawyer',
who turned out to be a much-maligned Chicano lawyer,
fighting for the miserable rights of other Chicanos who sought
his help.

They had met by accident in a bar in Aspen. Oscar captured

Hunter's spirit and drew him into the causes of his Chicano clients in East LA. One case in particular got Hunter started on his journalistic crusade with *Rolling Stone* – a magazine which had been launched in San Francisco in 1967 by Jann Wenner with $7,500 he had borrowed from the journalist Ralph Gleason, his mentor and dear friend. The story was not so easy but engaged Hunter in what became one of his substance-fuelled protest crusades against injustice.

Ruben Salazar, a radio announcer on a Hispanic station, was also a journalist with the *Los Angeles Times* and had been covering a riot in East LA with a radio reporter. The story goes that they had stopped for a beer in a bar on Whittier Boulevard. Somebody told the police that two men were in the bar, Ruben and his colleague, and one of them had a gun – an irresponsible thing to say under the circumstances. They were asked to come out with their hands up. When nothing happened, the police shot tear gas canisters in through the door. One of the canisters hit Salazar on the head and killed him. The police denied any involvement with the death and tried to blame it on the Chicano rioters, but there had been many witnesses, and so they were forced to admit that one of them had, indeed, fired the fatal shot, a fact later confirmed by the coroner's office. The Chicano community used Salazar's name to further their cause for equal rights and 'Remember Ruben Salazar' became a war cry, his name the symbol of their repression.

Let me say it here and now. For all Hunter's mindless self-indulgence, which is legendary and crude, he always impressed me with his blind, selfless urge to cut out the crony bestiality of modern society and the political calumny that scarred that era. He was, for God's sake, one of us. I believed him, was inspired by him and allowed him in his crusade to do what was necessary. He never let me down and as far as I know, when we were on that ride, whichever one it was, he got from me as good as he gave.

The Salazar case, which appeared in *Rolling Stone* on 29 April 1971, suggested that Hunter was not merely playing for effect, but believed that his writing was worth more than a one-off appearance in a new fancy rock'n'roll magazine. He legitimized the paper and people began to take it seriously as *the* magazine to watch and read. Since our lucky bull's-eye meeting in Kentucky, I was along for the ride, but I knew instinctively that my images, when I produced them, put Hunter's words into a truly legitimate scenario. I could visualize every nuance of meaning he wished to project. I knew it from day one and I knew it right up to the case of Lisl Auman who was imprisoned for life in 1997 for the shooting of a Denver police officer, even though she was handcuffed and in the back of a police car when the crime was committed. Hunter was outraged and fought for her release, which came to pass after his death. In that way he was a genuine crusader whose generosity knew no bounds.

The turgid Salazar investigation had been getting him down – he was a mess of nerves and sleepless nights, kicking and screaming, as was his way – a desperate, frustrated writer who really wanted to be a *proper* writer, not just a journalist. His one novel, *The Rum Diary*, had been a failed experiment and kicking and scratching was all that a trapped animal can do. He called the one man who could offer him help.

Oscar was forever surrounded by mean *gringo/gabacho*-hating friends who wanted no part of Hunter, except, perhaps, his head on a stick. In desperation, Hunter had kidnapped Oscar in a rented car and had driven him over to the Beverly Hills Hotel, away from Oscar's own environment and poured his heart out in the Polo Lounge. Although Oscar was sympathetic to his predicament, he was also headman to all those disparate souls and they needed their appointed leader. But Hunter needed Oscar more at this point.

Someone at *Sports Illustrated* had asked Hunter if he would be interested in going to Las Vegas to cover a motorcycle race called the Mint 400. He was interested and suggested that it just might be a good idea to get away from the madness of the Salazar murder implications and, at the same time, clear their minds by doing something completely mindless. So, he agreed to take on the 'Vegas thing' if he could take along his new assistant, Oscar. After the trip, which was supposed to be about the race, *Sports Illustrated* rejected the 2,500 words Hunter filed and refused to even pay his expenses.

But, at this time, Hunter's mind was somewhere else. He was holed up in a Ramada Inn in Arcadia, California, near the Pasadena Auditorium, right across from the Santa Anita racetrack, working on the Salazar piece. His days were spent sleeping, while his nights became a fuel-injected writing frenzy about Salazar, incorporating some of Oscar's straight-talking about the case. However, his mind was also freewheeling through a Vegas landscape of strange people, racing freaks and hack journalists. It was obvious that he was sick.

This was the best time to invent something weird and significant. Hunter called it 'a qualifier – the "essence" of what, for no particular reason (HAH!), "I've" decided to call Gonzo Journalism.' Noooo! Conveniently inaccurate. It was a strange chemistry that brought it about and Hunter had thought the Kentucky Derby piece a 'Goddamn failure' until it appeared in *Scanlan's* and a journalist friend from *The Boston Globe*, Bill Cardoso, wrote to him saying: 'Hey, man! That Derby piece was crazy!! It was pure GONZO!' And that was the very first time that Hunter, or I, had ever heard the word 'Gonzo'.

He picked it up immediately and made it his own, but, at the same time rationalized it by describing it as 'a style of "reporting" based on William Faulkner's idea that the best fiction is far truer than journalism' – and the best journalists have always known this.

His idea was to buy a fat, spiral-bound notebook and record everything as it happened and then get it published exactly – facsimile – without editing. He wanted it to feel like his mind and his eyes were functioning simultaneously, like a Cartier-Bresson photograph – no cropping – the entire negative, without the usual futzing about in a darkroom later. He hungered after a mind picture. He welded into one the talents of a master journalist, the eye of an artist photographer and the 'heavy balls' of an actor.

Only *Rolling Stone* seemed ready for an inspired accident

like this. His fired up resentment towards a fascist/Nixon government which was likely to get re-elected, coupled with the *Sports Illustrated* blank rejection and his 2,500-word 'experiment' all conspired to jam his mind full of the hate he needed to go wild in the most artificial city on the planet, Las Vegas; to break the law and leave behind a litany of felonies that would put anybody else behind bars. It was a last-ditch attempt to live up to the wild freedom of the sixties before convention and 'common sense' closed it down for ever. He called it 'conceptual schizophrenia', caught and crippled in that vain, academic limbo between 'journalism' and 'fiction', hoist by its own petard.

But Jann Wenner, who, like a Hogarthian wet nurse, kept *Rolling Stone* comatose on gin, vodka and drugs, seized the wildness and gave it its head. That was all Hunter needed, along with his wild 'Samoan' lawyer, drawn in by the progress of the 'story'.

The subject of Ruben Salazar was all but wrapped up and the craziness was now affordable. Nobody could overhear their plans as they plotted, in a top-down red beast of a red shark on the road to Las Vegas, a crazy plan and they would live to tell the tale.

Hunter called me, to tell me more: 'Oscar is a bit fucked up, by the way. He suffers from ulcers and self-doubt. And he doesn't have many clients . . . well, one actually, an actor who fell off my motorcycle and broke his leg . . . Yes, Ralph! It was because of me!'

'How unusual,' I replied, 'and then what?'

'One thing led to another and I asked him to accompany me on a journey to the Heart of the American Dream. I was going to ask you, but after that Rhode Island business, I reckoned you would have had enough. And I needed a lawyer — even a Samoan one.'

'I thought you said he was Hispanic.'

'Well, he is, Ralph, but for the sake of this story I have written Samoan sounds better. Anyway, what I really called you about was whether you would be up for doing some vicious drawings for it if I send you the manuscript?'

'Yeah, okay!' My heart pumped faster and I got butterflies in the stomach. 'Send it over!' I said with false bravado. 'I'll see what I can do. Who is it for?'

'This music magazine, *Rolling Stone*. They've never heard of you but I assured them that no one else could do what I want.'

'Then I had better not disappoint them then!' My confidence grew and I began to feel that at last I had found an outlet for all the pent-up traumas and mental turbulence I had suppressed over the last six months. I had been pretending that there was nothing there; things were going well and I was seeing my four children every Saturday.

It was as though I already knew the story. I had been there before. Not the same place, not the same story, not even in the same skin, but a shock of recognition from a suppressed well of personal experience and personal dread. An exciting resonance with something suicidal emerged and I settled down at my ink-stained drawing board in the back bay-window of our living room on the slightly raised first floor of a Georgian terraced house at 103, New King's Road, Fulham, London SW6. I dipped my steel pen — now a lethal weapon — into a blood-black cauldron of bile and began, accompanied by beer and brandy chasers, the therapeutic exercise of expunging from my mind all those trapped demons that lay in wait for their

mark of recognition, so that they might emerge blinking and grimacing into the harsh daylight of reality. I was there to give them life in whatever form they chose for themselves, like a theatre costume department handing out wigs, gelatine masks and rudimentary skin-tight costumes for each to play its role, as it saw fit. Then I did the same for *Fear and Loathing in Las Vegas Part II*.

One day, maybe, they will set a blue plaque into the wall outside that house which says: 'In this house in the summer of 1971, the artist Ralph Steadman (1936–2036), poured his soul out onto paper to liberate the evil demons from a manu-script by Hunter S. Thompson, entitled: *Fear and Loathing in Las Vegas. A Savage Journey to the Heart of the American Dream*. Res Ipsa Loquitor.'

I felt purged and better. It had been a psychological throw-up, a mental mess of half-remembered terror, agonizing flashes of an assassin's self-doubt, a half-healed scar, a screaming lifestyle expelled. I hardly drew them myself. I simply let them happen before my very eyes. I sent them off sometime in late September, informed my agent, Abner Stein, and resumed my work on *Alice through the Looking Glass*, a less haunted person than I had been for some time.

The concept was by Hunter but the pictures, drawn by me, augmented the crazy dimension he had hoped he could single-handedly create in his fat notebook. Following my own contribution, everything was done, processed and realized in two copies of *Rolling Stone* – Issues 95 and 96, 12 and 25 November 1971, respectively.

In the book which followed seamlessly in their wake, it reads, however:

Copyright © 1971 by Hunter S. Thompson.

All rights reserved under International and Pan-American Copyright Conventions. Published in the United States by Random House, Inc., New York, and simultaneously in Canada by Random House of Canada Limited, Toronto.

Fear and Loathing in Las Vegas by "Raoul Duke" first appeared in Rolling Stone magazine, issue 95, November 11, 1971, and 96, November 25, 1971.

Drawings by Ralph Steadman originally appeared in Rolling Stone issues 95 and 96, November 11 and 25, 1971, respectively, © by Straight Arrow Publishers, Inc. All rights reserved. Reprinted by permission.

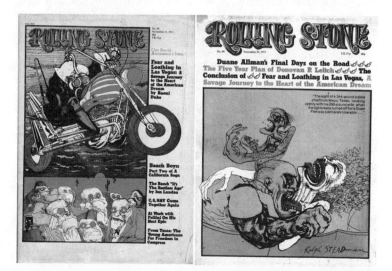

So, not a breath of copyright for me, Ralph Steadman. I didn't notice at the time, never read that page in the book, naturally, because the buzz was so great. Praise was overwhelming and chicanery was the last thing on my mind. The book was noticed mainly for the drawings and through the years, unknown to me, they were milked and used mercilessly. They instantly became iconic and an integral part of this great new book by Hunter S. Thompson, so far declaring himself a Doctor of Sophistry (i.e. 'anything that fits is okay by me', as he once said, explaining the definition).

I was being gently screwed and separated, ever so silently, from my own baby. It never occurred to me that my work would be subject to so much greedy frenzy. There was a use made of it that I can only describe as a hyena scrum. The worst was the journalistic scroff of pretending the 'wow man we dig what you do and we need it for our next issue, and fuck you because we'll use it anyway' attitude and I was too stupid and nice to say anything against such people. They were scum but I had not yet learned to know the difference. Many rock groups have gone through the same legalistic sponge, and they have learned the hard way, like I did.

I remember asking Hunter what had become of Oscar. I had met him a couple of times at Jerry's Bar just across the road from *Rolling Stone*'s San Francisco offices on 3rd Street, an old shunting rail-yard area and warehouse district of sand-blasted brick walls and submerged rail tracks. Oscar had appeared to me to be such a gentle, private soul who told me of the poor young kids who had got themselves a police record

simply for smoking pot. He said it was unfair to mark these kids with what was not a criminal act. He told me that he defended them as a crusade of principle. He struck me as a genuine and humane man.

When I asked him, Hunter didn't say anything specific about Oscar's fate — just pointed to his solar plexus and pushed, suggesting that he had been shot in some drug-running scam or maybe — and I prefer this version — he had become involved in a mini-revolution in Puerto Rico and had been shot fighting for the rights of the people he had spent much of his life defending. In truth, Hunter preferred I didn't ask.

About three weeks after I sent off the pictures, I received a letter from Hunter.

> OWL FARM. Oct 19 '71
> Dear Ralph . . .
> I've tried to call you at about 19 fucking numbers and everywhere I call, they curse your name. Then I got your card from the south of France, so at least I know you're alive.
> What I wanted to tell you is that the whole Rolling Stone office in SF flipped, freaked & went crazy when they opened the 'London packet' and saw your drawings for Vegas I. All I had to say was, 'Shit, that's my man. Who else could have done it?'
> Who else, indeed? They were fucking beautiful. I told Wenner right off, that nobody else could possibly catch the madness of this story & that I refused to let anyone else illustrate it . . . but, Jesus! I was overwhelmed when I saw the shit. Fantastic . . . & I'm awaiting Velox copies of your drawings for Vegas II which Wenner says are even better than the first batch.
> Your snout-bike will be on the cover. We dominate the whole issue . . . And in the

meantime I went to NY & sold the Vegas horror
as a book to Random House, and I've told my
editor (Jim Silberman) that we must have your
drawings for the book version. He's agreed,
but he's nervous about the cost. Neither
illustrations nor photos seem to help a book
sell in this stinking country; all they do is
jack up the retail, per copy price, and that's
naturally a critical matter, to everyone involved
- including me, and ultimately even you.

Etc., etc., etc. . . . the letter went on quoting all the reasons
I have heard since, *ad nauseam*, publishing economics began,
and Hunter was always as cautious as the rest. 'But,' he
continued, 'I definitely think we're on top of this one, unless
your agent runs amok and demands something like ten grand
— which would kill it all (the drawings: not the book — it's
already sold and heading into print. Incredible. I still can't
believe they would pay me for writing this mad gibberish).'
What actually happened was that *Rolling Stone* paid me
fifteen hundred dollars for the use of all the drawings — about
twenty-four of them — and then offered to buy the originals
from me, which my agent urged 'was a good move!' He sold
the whole damn treasure-trove to Jann Wenner for the princely
sum of sixty dollars per drawing. I rue the day I let him
convince me.
Random House paid me five hundred dollars for the use
of the drawings in the 1971 hardback edition, plus another two
hundred and fifty dollars for their subsequent use, times
millions of copies worldwide for the soft cover version. No
royalties were forthcoming to me until Octavia Wiseman, who
worked for my agent at the time, managed to convince the
publishers that I did indeed deserve some residual compensa-
tion and financial interest in the book's continued success as
the drawings still assume an important role in its performance. It has been tentatively and grudgingly acknowledged and
I have received one per cent interest in something but am not
yet quite sure what it is an interest in.

This has been a lesson hard learned and still they catch me out. It was the drawings that alerted potential readers in the first place that this was something to take note of. Drawings, or 'illustrations', as they are miserably called, can be the means of energizing the life of a text. Where is *Winnie the Pooh* without its illustrations? Where is *Fear and Loathing in Las Vegas* without its Gonzo drawings? Impossible to say now, in retrospect, and I wonder still, but I wouldn't have missed the trip for the world.

HUNTER GOES TO WASHINGTON
1970–72

*From here to modernity . . . The New Journalism
comes of age . . . Hunter joins the political process . . .
Freak Power . . . It never really happened anyway
. . . Sweating it out in Miami*

Ralph
    I have to be in Washington DC (for a year,
leasing a house there & the whole trip) by
Nov 1st '71 Which means total madness here —
packing up everything and renting this house
for a year. I'll be writing a column for
Rolling Stone, every issue, until the Nov '72
presidential election.

Hunter invited me to join him there and then, thinking,
no doubt, of the *Fear and Loathing* drawings, added:

    Shit, maybe you should never go anywhere.
Just do all your work at home, from mailed-in
manuscripts . . . god knows, these Vegas drawings are a hell of a lot better, overall, than
the Kentucky Derby stuff. Very weird — especially to me, because I must get personally
involved in a scene, in order to write it.

He had rented a house facing Rock Creek Park and invited
me to come and stay, adding that my only real problem would
be the plane fare. It was a lot of other things as well, at the

time, but no matter. *Rolling Stone* was still feeling its way in this new genre and hiring an artist from England was not a priority. Warren Hinckle, the genius editor of *Scanlan's*, had long gone from American journalism, having found a 20,000-dollar-a-year job as a 'management consultant' for a chain of supermarkets. He brought them to the brink of bankruptcy within two months.

Hunter was still pressing a lawsuit for 3,600 dollars against *Scanlan's* in the New York courts, if only, he said, to deprive Warren's partner, Sidney Zion, of the money. There was genuine acrimony between them and talk of ripping off of nuts in a New York elevator, but only an embarrassing scuffle ensued. The lift operator, not your regular little guy, but a six foot four one-time wrestler, kept them apart up to the twentieth floor while they screamed obscenities at each other. I reckon they owed me too, but from where I was sitting, recompense for losing my pants and shirts and nearly my life in Newport, Rhode Island, seemed like small beer compared to the wild experience which I considered something far more precious.

Nineteen seventy-two was the year I finished the drawings for *Alice through the Looking Glass*, I had started to do fortnightly drawings for *Rolling Stone*'s 'World News Round Up' and was planning 'something ecological'. I had hoped to involve Hunter's take on it and lash a book together that would at least be a book of excess, contradicting everything anyone was saying about the environment. Nothing was really serious in this fashionable new obsession and in retrospect all it has ever achieved is a ban on smoking in bars and restaurants. Even today, oil reigns supreme. Cars are exempt, particularly those four-wheel-drive tanks that suck up gas like submersible flood pumps and throw up like dragons. They are usually driven by self-righteous little mothers, who are obsessed with getting their wretched offspring to school on time from about one hundred yards away from their homes. Like their children, they can hardly see over the dashboard and like their children they park as though they are in a ploughed field. They like to put a sign in the back window which says: 'CHILD ON BOARD'. (Huh! Which one? Driver or passenger?) I always want to add: 'IDIOT DRIVING THIS CAR'. That I might have written it with a big black felt-tip pen on the back window next to the sign, I don't want to say, for fear of vicious recrimination.

They are the types who rail against smokers, but they are guilty for the toxic pollution on this planet along with all the rest of us to a greater or lesser degree. But smokers, after all this time, finally got the bum's rush and are, along with winos, travellers, drug-pushers, beggars and pimps, the most anti-social vermin on the perimeters of society.

Anyway, Hunter responded to my earnest request:

```
Ralph,
    You illiterate bastard. I had to send your
letter to an interpreter to figure out what
kind of 'ecology book' you were talking about.
Needless to say, I'd like to work with you
on it - but anything involving a lengthy text
is out of the question for at least the next
six months. I have a '72 Campaign book due
on Jan. 1, and after that another book for
Random House. But if you can get by with a
series of captions, instead of a full-length
text, I can probably manage to do it - providing
you can arrange to have me flown over to
London for at least two weeks, so we can lash
our ideas together . . . To that end - etc
. . . Avoid . . . AR and SZ . . . must have
been weaned on the same ugly tit . . . though
. . . with the possible exception of a huge
advance in the form of gold bullion. Etc
. . . but I like the idea of your twisted
ecology book, so why not send me some details?
Or at least one. OK for now. HST
```

I sent him rough sketches of oily, jammed machinery that had stuck at high revs and was pouring out thick black smoke. There was only an 'on' switch. The earth on a life-support system, a human foetus inside the womb of the world, attached by the umbilical to the inside core wall of the earth's crust. The impossibility of our appetites, even then, only thirty-five years ago. What's that in the history of the world? A dust particle? A Saudi Arabian King holding the world as a testicle, which was attached to a petrol-pump dick, and a caption which stated: 'Right now, it's a one-balled world.' Another − a long line of cars on a desert landscape, not unlike Las Vegas. Cadavers sat in the driving seats or prostrate on the ground, holding petrol cans, helplessly dying, or dead, as they had tried to find petrol. Up in front as a huge roadside billboard, a poster of apology: 'SORRY! NO MORE PETROL, EVER AGAIN.' Hunter liked these ideas and his crusading spirit was whetted. However, as with many of our enthusiastic discussions by letter or phone, nothing was going to come of it, mainly because of personal agendas, private ambitions and in Hunter's case, a solemn promise to sell his soul to *Rolling Stone* and then with a twinkle in his eye − buy it back.

'Fuck them, Ralph! Even the Devil has to pay!'

'I didn't know you had read Goethe, or listened to Gounod's musical version,' I said.

'Fuck them too, Ralph! They never read my stuff and that's all we need to know. I have tapped into a rich, greedy vein and I will milk it like a terminal heroin addict.'

'I suppose you think I can provide suitable visual stimulation for all of this − Me! A simple artist!'

'Yes! You, Ralph, but not simple − just cunning and devious. No one will ever suspect that you are the Devil's disciple, or, for that matter, that I am the Devil.'

I responded to his claim with derision. 'Hah! You are not bad enough to be the Devil. I am the Devil! You are just a fucking writer. I am an artist, I trade in uncertainty and superstition and cant. I invent dark visions of impossible situations that can never be resolved. My thoughts are only suffered but they are never resolved. You simply use the language of everyday events that find everyday solutions.'

'We are all artists, Ralph, but I am the one with ready access to your sick visions. To you it may be a few words, but to me words represent a minefield of possibilities, a pig-fucking suction machine on the grease nipples of life.'

Nicely put, I thought, but did it come from him – or the Devil?

While we had been in Newport for the America's Cup, Hunter had begun to make plans to run for Sheriff of Aspen. In 1970, the incumbent was a Democrat called Carrol Whitmire, who upheld the law over a laid-back bunch of refugees from the sixties – hippies, dreamers, artists, musicians and a diehard bunch of old-time Republicans who had grown up and grown old in the place and didn't want change – a perfect spot in the Rockies for gentle corruption to flourish. Life was easy, steeped in folklore – and remote. To get from Denver to this former silver-mining town, you can either drive up on the 'main road', Highway 70 West, through a hot spa town called Glenwood Springs, or take the interesting mountainous road through an older and more historic silver-mining town called Leadville, boasting its own Tabor Opera House and visited in the 1880s by Oscar Wilde, as a stopover on his American book tour on behalf of the Aesthetic Movement.

The roads were just opening up and in the spring and summer season you could drive over a 12,500-feet-high pass known as Independence – the great divide on the Rockies land-mass stretching from Canada down to the Gulf of Mexico. It was here that Hunter had settled and had been forced to make his once in a lifetime bid to get into politics and 'beat the opposition like gongs'.

When he settled near Aspen in Woody Creek in the late sixties with his then wife, Sandy, and their son, Juan, I have to imagine that he felt that he had found his Garden of Eden and had fallen in love not only with the free lifestyle of the town but its natural beauty as well. He had had his first serious run-in with the power and might of corrupt politics in Chicago in 1968, when he accidentally got involved in the political ruckus known as the Chicago Riots and got beat on. He told me that was when he got politicized. When we first met in Kentucky in May 1970, politics were not on the agenda, or, at least, he didn't mention them to me. However, a couple of months after our meeting, after I had returned to England, and after the publication of the first piece of Gonzo reportage, which set the hallmark tone of aggressive imagery, Hunter contacted me and asked if I could possibly do some sick drawings of the opponents he was up against in his campaign to take on the local law department. He sent me three black and white photos of the most unsavoury bunch of opponents you wouldn't expect to find anywhere but at a Nazi rally of the 1930s.

I did three drawings based on a theme which stated: 'Don't vote until you see the whites of their eyes; Vote Primeval Slime (if that's what you want); and Take another look at your Sheriff and vote before he shoots'. I felt that Hunter needed an uncompromising message to back his campaign. He was running on what he chose to call the 'Freak Power' ticket. It was his all-out effort to change Aspen 'the Democratic middle way', back into an honest and decent town. It had been getting sleazy. He issued a campaign manifesto which called for a scheme to rip up all the streets with jack-hammers and melt the blocks down to build a huge auto-parking complex and parking lot on the outskirts of town, preferably between a sewage plant and a new shopping mall called McBride's. He would establish a central refuse and waste disposal tip in the same area in memory of Mrs Walter Paepke, who had sold the land for development. Then all the streets would be restored to grassland and footpaths. The only automobiles allowed in the town would be for deliveries and other vital services like the Post Office, these limited to a network of 'delivery alleys'. A fleet of bicycles would be there for public use and would be maintained by the police.

Hunter also promised to change the name of Aspen to Fat City. Aspen as a name in the postal records would no longer exist and road signs and maps would have to follow suit. The huge Corporations that were beginning to move into the area, rape the land, buy low, sell high and then move on, would be struck a savage blow and greed-heads and ski resort developers would be discouraged. As he gently pointed out, these human jackals could no longer capitalize on the name of Aspen and should be 'fucked, broken and driven across the land'. Then he declared that he would hold a Drug Tribunal on a weekly basis on the lawn in front of the Sheriff's Office to 'listen to bad drug complaints, punish dishonest drug dealers by putting them in stocks set up in the open'. He declared that any drug worth taking shouldn't be paid for anyway and serious exploiters of drug-dealing would be dealt with under the new law. He would listen to any complaints at any time of the day and night regarding bad drugs and explained that all sheriffs in the state of Colorado are legally responsible for enforcing state laws, even though Sheriff Thompson may disagree with that law. Only legislation through due process could change that. He felt that the burgeoning drug culture in the area would establish a human ambience for those who become a part of it. The law would make things very ugly for those who came annually to profiteer in drugs or for that matter anything in any field. And finally he declared that the Sheriff and his Deputies would never be armed in public. A pistol grip Mace-bomb would be more than enough to deal with any disturbance, thus making Fat City a very safe place to live for all law-abiding residents.

It was a very Freak Power platform to fly, but he generated a massive young vote who considered the statement a constitution in its own right and a worthy upholder of the First Amendment for the right to free speech. Hunter secured 34 per cent of the overall vote and failed by a whisker. It was a magnificent rebuttal to those who preferred that a community be run on corruption, blackmail, bullying and deceit.

A year later and another election was approaching. The 1972 Republican and Democratic conventions would both be lived out in Miami, but not by me.

'Do you want to come to Miami, Ralph? Jann has agreed to you being there to capture the sheer mad enthusiasm of a convention from the floor.'

'From the floor? How do you mean from the floor?'

'We've got credentials this time. *Bona fide* documents of

authority. We will be rubbing shoulders with the great and the good. What about it?'

'Why didn't Jann ring me himself?'

'He thought it would be better coming from me. Working partners.'

'Tickets there and back? Expenses? Hotels? Proper ones — and insurance?'

'Of course, Ralph! This could be big and we'll dominate the magazine again just like we did for *Fear and Loathing*!'

'I'll talk to Anna about it and get back to you.'

'Okay, Ralph but don't wait too long. It's all wrapped up — both conventions.'

I was very excited and after many assurances and enthusi-a-stic speculations, we agreed that I would go alone and maybe come home between the two conventions, travelling back for the Republican bash in late August.

When I took off for this assignment, however, I didn't know that I was a sacrificial lamb for an American slaughter.

The humidity and heat when I arrived were augmented by too much in-flight booze. Flying was still an adventure and I hadn't the slightest idea that too much of anything can make one troublesome. I was mellow at first, watching the in-flight movie, but by the time I stepped off the plane I was full of Dutch arrogance. I was apprehensive too. Working with Hunter was always a challenge and he had infected me with an aggres-sive edge — or sharpened the one that I already had. My polit-ical drawings have always been more of a weapon than an item to entertain within the pages of our newspapers. I believed I was on a savage crusade and I believed that the work would have a similar devastating effect to the drawings done by George Grosz and Otto Dix during the Weimar Republic years in Germany after World War I, when Fascism was taking root. I simply transplanted some of those unforgiving visual statements into our own political scene of the sixties which I felt had adopted some of those pre-war sentiments. My attitudes suited the bolshie antics of the satire boom which was the fashionable humour at the time. It had become an industry and any other kind of humour was considered as redundant as pre-war music hall.

My stance was reinforced by memories of living much of my infant days watching my mother knit to steady her nerves, as we sat out the bombing raids around Liverpool and Wallasey in an Anderson shelter. Mornings after the raids were for me a surrealist scramble through the neighbourhood rubble of houses obliterated during the night, searching for pieces of shrapnel, the hard, melted slivers of metal that had once been part of a German bomb. During the intense heat generated by 500 lbs of high explosive in a split second, the components of such devices would melt and set again into whatever shape they were forming as they seared at high velocity through the air or suddenly came into contact with a solid or living object. This would not concern a child, if a child would be at all aware of what these by-products represented. Nobody came rushing out to alert us to the possible dangers. Occasionally, a sign was erected by an air raid warden which said: 'DANGER. UN-EXPLODED BOMB', but no further warning was heard until the voluntary services – usually the Home Guard – had done their job. Anyway, any further warning would have been superfluous to requirements.

These hideous, white-hot missiles would be nothing more than valuable items the next day for playground currency. Three lumps of shrapnel of unusual shape and size could get you a Hornby clockwork train in exchange and six would get you the rails as well. We may well have been dealing in the grizzly memorabilia of what had once been 139, Poulton Road, but we wouldn't understand the symbolism enshrined in these objects as we played with them or passed them on for profit. It may still be happening today in Fallujah and Basra or anywhere for that matter where the bomb is a daily occurrence.

I attended a local nursery school at the time and I wore an apron which sported an appliquéd elephant and I remember that my mum and dad used to joke with me and their friends that an elephant never forgets.

Hunter would never have had such memories because a blitz was unheard of in the United States of America until the nineties, when the results of what they remotely called their 'foreign policy' began to filter back home in demonic ways. Hunter loved high explosives and weapons of many kinds and nurtured an arsenal of them that unnerved me, though I did come to accept them as a part of the package.

I hated Miami.

'My Miami hotel room looked like a place one would go to commit suicide – or just die,' I wrote at the time. I just wanted to get back home. Collect all the material I needed and then flee. 'People in America don't want peace,' I wrote. 'They just enjoy a screaming lifestyle. American politics has lost the grass roots so necessary to motivate people and not just make them rant and rave like child lunatics. We shall overcome – Huh! broken, hoping faces – a melting pot of humanity without a foundation – hope? – not a hope in hell – the sadness and defeat of the great American Dream is written all over the middle-American plastic smile – Miami was/is an alligator swamp – it is time it was redeveloped.' I was scribbling down all these thoughts and I hadn't met Hunter yet.

We got together at an open-air café near the beach for breakfast. I was feeling nauseous but I always eat a good break-fast. I was already expressing my doubts about being in Miami but we agreed to attend a rally that night in the Convention Hall to hear George McGovern give his speech accepting the nomination of the Democratic Party. 'The nicest thing I have seen so far' – I wrote – 'a blue tit eating from the table we were sitting at.'

We agreed to meet later that evening and stumbled into the hall past security. It gave me the idea of seeing this heaving mass of electrified madness through the stranglehold grip of one of the security guard's armpits. That would be my preferred satiric stance. I spent the rest of that day going in and out of monumental hotels like the Fontainebleau and the Doral just to get cool. The air-conditioning was just as oppres-sive as the humidity outside. I was beginning to develop a fever and I didn't belong.

I dragged myself over to Flamingo Park to see the protesters, the Vietnam vets in wheelchairs – the survivors of another war gone wrong. Don't they all! Tents had been erected and Portaloos were scattered everywhere, in direct contrast to the luxury of the hotels containing the committed delegates and their spangled wives and mistresses.

'Four More Years' was the Republican war-cry and warning and four years was what they would have got had it not been for Watergate and the dirty tricks of the incumbent administration, but these Democrats were equally decadent! 'There is about as much politics in a convention hall as in a detergent ad,' I wrote. Vicious gaiety broke my spirit — it was like trying to smile in a gas chamber. 'Authority is the mask of violence,' I wrote and continued to write for the brief twenty-four hours I stayed in the place.

In the hall, people were crazy with ecstatic grief, the kind of grinning grief you know is happening as you watch a multitude of fully paid-up Democrats bravely live out the last moments of their lives as they had known them, trying to believe that the next four years was theirs.

I thought they were like Christians who had staggered into an arena surrounded by Romans who were there for the sport and they were only waiting for a brave 'Hurrah!' from their chosen leader, who would deliver them into eternal life hereafter. To their credit they reminded me of alligators (Christians with teeth) galumphing in Oxford boater hats, grimacing and

cheering when their leader appeared and declared from the podium: 'Come Home, America!' The cheers that followed were more of a dying scream than a confident reaffirmation of what could be good about America – if only things were different and not so fractured by a war that even the Republicans didn't want. It was a brave stance but not one that was going to break the stranglehold that the Republicans instinctively knew they still had. It was coming – but not yet.

'As Maine goes – so goes Vermont,' I wrote. 'And Massachusetts, Idaho, Illinois, Missouri, Montana, Oregon.' At the time most states still couldn't bear the change, the upheaval, in the midst of an unresolved war and so would settle for the devil they knew. For every Democrat there are two closet Republicans complaining about what they are about to vote for again. I'm not an expert on American politics or, for that matter, English politics, but I have a sixth sense about people and their 'terminal anxiety' – that last-minute decision in the privacy of a polling booth. Why change in mid-stream? Better to see this one out in case they have a master plan. Sure, I can scream for one leader, and support him/her, but not now, not yet. Which is exactly what happened when George W. Bush, a certified halfwit, got re-elected in 2004. That was the thinking process that really killed Hunter S. Thompson. That broke his spirit. That was it. Not even Gonzo could cut through that Gordian knot.

We were maimed by a night of wild, crazy enthusiasm for something we knew in our hearts was not going to be and neither was it going to hold me here. I had decided to tell Hunter at a late breakfast we had agreed to enjoy together that I was going home. I had seen all I needed to see, and as for the Republican Convention a couple of weeks later, I had all the material I needed to do that, too. Besides I could fill in the gaps in England, where the proceedings would be televised.

'Holy Shit! Jann isn't going to like this. He's flown you over and put his faith in us both to deliver ball-breaking coverage – and now you tell me this. What can I tell him?'

'Tell him the truth. Tell him I got sick, both physically and mentally. Your politics crushed my spirit.' Tim Crouse, a

Rolling Stone editor, was with us and he remained silent. It was bad karma. He knew Jann better than me and realized what a smack in the mouth this would be. But I was adamant. I couldn't wait to get to the airport and fly home.

But Hunter sensed something strong would be the result. Had I ever let him down? No, true. Then trust me on this one. I am sick and when I get home I will recover and get to work. My reason for leaving was nausea.

When I got to the airport, I had time on my hands. After the usual tour of the souvenir shops, the bars and duty free enticements I was in for a boring few hours of waiting. Then I came across an arcade of fruit machines, fortune-telling, card-delivering, macho-proving slot offers. 'Are you sexy?'; 'Are you really a man – or a woman?'; 'What weight do you think you ought to be and the reason why?'; 'Drive a racing car all by yourself and win a Formula One car'; 'Assassinate a President and check your score'; 'Make your own dark-green alligator and take home an authentic souvenir from Miami'. What? 'Simply place 25 cents in the slot and make your own alligator memento.'

I had a couple of quarter pieces, so I slipped one into the machine. There were sucking noises and sounds of flatulent bubbling and farting, as, within, the machine was melting rubberized plastic to the correct temperature, mixing it and depositing it into a mould. Then there was silence and another fart, silence again and then a trundling thump, like a Coke bottle being dispensed, as something fell into the receptacle at the base of the machine. I opened the flap and reached inside for what was an object – warm and rubbery. I raised it carefully into the light and saw that I was holding a well-formed model of an alligator and the fact that it was still warm, compliant and khaki brown made it all the more hideous. It was the perfect memento. It became my talisman, my *bête noire*, my *raison d'être*, the object that remained on my drawing-board throughout the period I spent tied to it in Fulham, pouring forth all my mixed emotions and my contempt for everything American at the time.

Within the space of three weeks all was done. I dutifully

sent the work off with the courier supplied by *Rolling Stone* and waited.

The first broadside appeared in issue number 115, 17 August 1972, spread all over. It was working and my apparent timidity was overlooked. The second issue, number 118, 28 September 1972, confirmed, as far as Hunter was concerned, that the whole shebang was a preconceived re-elect Nixon circus. By endorsing losers like Goldwater and Spiro Agnew, who at the time was Vice-President and one of the most hated men in politics, Nixon was showing that he didn't give a damn. Teddy Kennedy could win for the Democrats and that would suit the diehards, who would be prepared to wait four more years. It was a sham and a gamble. Spiro Agnew was never going to make it, so Nixon was, in effect, trading him off for an assured second term. That, in a nut, is politics and as far as I can ascertain, still is. It is all about strategy and screw the needs of an electorate.

In spite of my apparent falter and Hunter's fear that my contribution would add up to nought, the partnership worked and we were a working outfit. I had divined, quite by chance or sickly instinct, just what was happening in Miami in the year of '72.

WATERGATE FOLLIES
July 1973

Assaulting Sam Ervin in the Caucus Room . . . Singing
Masters of War with George McGovern . . . Fishing
with Juan in the Roaring Fork River

'It seems such a long time ago that we arrived here — two
and a half weeks ago,' Anna had written in her diary, 'after
an inhuman flight across the western hemisphere.'

We arrived in San Francisco on 25 July, exhausted after a
fourteen-hour flight. I was wondering what in hell I could do
to encapsulate a sense of America as though I was a clone of
Hunter — a part of America that Hunter could only capture
in words. We were expecting Jann Wenner to meet us at the
airport — but why should he? He was a busy man. Even his
minions were busy.

It was 9 p.m. local time and 5 a.m. in London. So, our
time-frames were completely askew. We took a taxi into town
and checked into a folksy-looking place downtown, the Mark
Twain Hotel — a modest place, the desk girl told us, and indeed
it was. Architecture in the style of the 1930s, homely with a
breakfast shop next door that sold beer. We bought hamburgers
and beer and retired to our room to sleep. Everything tasted
good because everything does at 4 a.m. when your body says
it's 5 p.m. yesterday.

We switched on the television and ate our little supper in
bed. It's what makes travelling so delicious. A sense of other-
worldliness you can only achieve when you climb on board
and go somewhere else. It is the real reason I like travelling.
We slept well, too, even though we woke up at four in the
morning, our bodies being unused to the time change.

The problem was that we didn't have Jann's home address,

so we decided to go to the office. It was a hot day and we decided to walk there. We left the stores and shops and found that Third Street was an evolving thoroughfare of new blocks, old buildings and demolished lots. Jann's office impressed us with its size and taste — bare brick walls, lots of woodwork and the number of people working there obviously reflecting the growing prosperity of *Rolling Stone*. My *Fear and Loathing* drawings were framed along one wall, plus some others that Jann seemed to have appropriated.

From the office we took a cab to Jann's house. It was a large Victorian colonial-style building standing high up from the road. There was a little boy standing on the front steps — Juan, Hunter's son. We very quickly took him under our wing and I made him a bow and arrow from bits of wood he found in the basement. It was beautiful. I determined to make another two when we got back home for Theo and Henry, my own sons.

Juan was a strange, funny boy — introspective, which is not surprising, given who his father was. He left little trails of destruction wherever he went. He had a savage curiosity which I believe hid his shyness and he avoided direct confrontation. He was always investigating things, and doing accidental damage. I watched him closely from time to time. I once saw him looking at one of Jann's wife's plants in the back yard, stroking the smooth stems. One of them snapped in his fingers. 'Hmm!' he muttered to himself. 'We could have done without that.' I turned away and pretended not to notice, because I wanted to laugh. 'Let's all go downtown and have some breakfast,' I suggested. 'Give us something to do, eh?' Juan was all for it.

The next day we took a trip to the office to look around and meet some of the staff. Everyone was hellishly young and enthusiastic. It was an open-plan office, except for Jann's, which was partitioned off in the corner. As a boss, I figured, he was probably ruthless. Everybody worked long hours and firing and hiring was a frequent occurrence. But when things turned out well, he became as excited as a child.

We got back from the office and, on entering the living

room, I stopped dead in my tracks. Paul Simon, the singer, was sitting on the settee talking to Janie, Jann's wife. I was completely taken aback, as I was and still am a huge fan. I got a bit tongue-tied and he simply said: 'Aw, forget it. Fame freaks everybody out,' which I thought a little too self-inflating but I guess he was trying to put me at ease and I guess they are gods when they are at the top and are treated thus. He was a small man, friendly but quite self-centred. He has what in England we call a cocky walk and, later, I studied him intensely from behind.

We all went out to dinner, Juan too, who fell asleep at the table with a carrot in his mouth. When we got back to the house he flaked out and we went to a record convention at the Fairmont Hotel. Paul wanted to change first, so we went to his hotel. He was wearing jeans and a baseball cap and the doorman said to him suspiciously: 'Are you staying here?'

I went to the Fairmont the following night, the last night of the convention. Art Garfunkel sang. It was great. He has such a pure voice. He seemed a little awkward singing, not sure what to do with his hands. Then on came Sly and the Family Stone, a black group I hadn't heard of before. They seemed to consist of two generations, including a young protégé doing a Jimmy Hendrix impersonation.

Jann told us to go home in a cab and they would follow. He had invited some people back. Five people arrived but no Jann or Janie. It felt very strange chatting to these people as we waited for a host who never came.

The following day we decided it was time to explore the city. So we took Juan with us, intending to take a cable car downtown. This had become quite a big thing because we had tried to get on cable cars before but they had always been too crowded. We finally got on one but were told to get off and wait for another one. We waited for a few minutes and suddenly a car pulled up opposite and the driver shouted: 'Juan! Juan! You little bastard!' It was John Clancy, lawyer friend of Hunter and tax advisor to Jann too. He had pockmarked skin and a wild look. He had been up all night taking cocaine and it had given him a bad hangover. He wanted to show us the city. He took us first to his house, which overlooks the bay. It was in a terrible state. The first thing he did was to go to his desk and take a 'snort,' leaving the front door wide open. He took us to see Boz Scaggs practising in a basement recording studio from which we could see the Golden Gate Bridge and mists swirling about the bay. John showed us some weird bars and Seal Rock Inn, one of Hunter's favourite haunts.

He also took us to an English pub, where I drank Watney's Red. An English boy heard our accents and came over to talk and John Clancy became quite violent, telling the poor bloke to shove off. I met John on a later occasion when I came to America alone and he showed me a darker underbelly of San Francisco night-life.

I recall, that time, going on board a merchant ship to be introduced to the crew. It was weird and I remember shooting guns out of his living room window into the night. It was a stupid thing to do. We must have been crazy-drunk but I do remember worrying whether we had hit anybody. I was there to cover the Patty Hearst trial on my own for *Rolling Stone*. Well, not exactly on my own. *Rolling Stone* had organized for me to meet Joan Didion in the bar downstairs and maybe team up with her to do the story with my drawings. We did not hit it off, somehow, and we agreed to work alone. It was all a bit *Alice in Wonderland* and I treated Patty exactly like that. It was the first time I had worked in America without Hunter.

The next time I saw John Clancy was March 2005 at Hunter's funeral and then again at the Memorial Blast-Off in Woody Creek the following August. He reminded me then of a sleazy English country gentleman. He was wearing a sporting navy-blue English blazer and a cravat, as though he were at Cowes for a spot of yachting. All he needed was a monocle and the picture would be complete. Two weeks later I heard the news that he had been killed in a car wreck. I was shocked by that, coming so soon after Hunter's Memorial Blast-Off and a sneaking thought crossed my mind. Hunter's violent death *had* put people in odd mindsets, but we'll never know. It simply added to the depressing mood that has descended on anyone who knew him well. Aw, Hell! I am digressing. I was with

Anna in San Francisco having a great time and the next morning, a Monday, we were off to Washington to cover the Watergate hearings.

WASHINGTON

As we stepped out of the airport building, the air was hot and heavy and seemed to have its own special smell. We took a cab to the Hilton, where Hunter had booked us in. It was a cheap-looking hotel but in its vulgarity it gave us a bit of freedom.

The nice thing about the Hilton was the swimming pool. Our day usually started with a swim followed by drinks by the pool, waiting for Hunter to emerge. He never got up before midday and had usually been up most of the night. When he took off his sunglasses, his eyes would be red all over and swollen with tiredness. While we waited for him, we would talk to Sandy, his wife. She was very thin, almost brittle, with long, blonde hair. She doted on Hunter and had lived with him through the Haight-Ashbury summers of love. She said they'd been madly in love for thirteen years and that this was the first time for a long while that she had left Owl Farm and travelled with him.

I had written some Watergate notes at the time: 'They were all lawyers or liars. All the bad guys go to law school.

Kalmbach, Ehrlichman, Haldeman, Nixon, Segretti. Loyalty was the name of the charade. Erlichman – he leaves no more blood on the floor than he has to. There is a type of evil here that has become okay. The term "evil", sir, is relative. I don't quite recall, sir . . . No one wanted to perjure himself. If we were in Germany in the old days, you'd be a lampshade and I'd be smiling.' Then I mused: 'Experts disagree as to whether Nixon's complex was the result of an early manifestation of Parkinson's disease, or neurosyphilis caught during his student days. It has hallucinatory effects on the victim, giving him a sense of grandeur, a hatred of humanity, near-impotence, and, finally, premature senility. The second possibility, however, has been ruled out on the grounds that at the time Mr Nixon was a student, it would have been impossible for him to contract such a disease, owing to his pathological fear of women, unless it was perhaps accidentally picked up from a lavatory seat. He died senile at the age of eighty-one in an anti-environment shelter at San Clemente, seated in a deck chair before a sun-lamp wearing only a pair of jackboots.'

It was chilling to realize that the entire administration was rotten to the core. They very nearly got away with it, until witness Alex Butterfield, who had been Nixon's former chief of internal security at the White House, mentioned the tapes. Tapes? What tapes? It was an innocent slip of the tongue. Nixon had been secretly taping everything that was said in the Oval Office by everybody since he had become President. The game was up.

John Dean, Nixon's lawyer, agreed to cooperate and tell all. He told me that the weekend before the Monday that he was due to testify, he had to go into hiding for fear of reprisals and he hid all his papers under the water tank in the loft of their house in Beverly Hills where he lived with his wife Mo. She was the only other person to know where they were. John and I worked together at the Kansas City Republican Convention in 1976 for *Rolling Stone*, and became good friends and the set of drawings for that Convention were done in his house.

*

One evening, Hunter told us he had invited former Democratic Presidential candidate George McGovern to have dinner with us at a Mexican restaurant. We found McGovern waiting in his car outside the Hilton, reading the *Washington Post*. He was a very polite, sympathetic man. At first the conversation was all about politics and Watergate. Then a young woman arrived who had campaigned for McGovern. She had a thirteen-year-old son who was smoking pot and she was worried about it. McGovern had invited her along so that she could talk to Hunter about it.

'Good man!' Hunter reassured her. 'He's got his priorities right. He can rape women later!' It was this direct approach, his wild candour, that also got my son off sniffing glue in his teens. He made us laugh.

'Sing "Masters of War", Ralph! You know "Masters of War".'

I had borrowed a guitar from a restaurant guitarist who thought I wanted him to play 'Granada', but I wanted to play the guitar myself.

'I do know it, Hunter, but I can just remember the chords.'

'Fuck the chords, Ralph, just sing the song and beat the guitar like an immigrant worker!'

Hunter was desperate for a war cry and this was Washington 1973. It was weird. It had to be. Rarely had such an unlikely group of people sat around a table in some downtown restaurant whose name I forget. I'm not paid to remember things like that. That's why I am not a writer. I don't remember names, but I remember occasions.

George McGovern was sitting opposite me. I certainly remember *his* name. He was Hunter's friend. George supported daycare centres and wage controls and is the only politician I can think of who was concerned about the environment as early as the environmentalists.

I had to be vetted before we could have dinner. That is standard. No problem. Was I okay? Could I be trusted? My drawings were weird to American eyes and strangely European. Was I a commie, perhaps, or even some flaky fanatic? Funny thing about Americans. They are the first to adopt weird lifestyles and radical views but they are the most conservative race on earth. It's a wafer-thin veneer that covers their pioneer, wagon train conformism. Their quintessential straightness cracks under pressure. They have to know where you stand to feel comfortable.

Anna was also at the table. I don't remember if George's wife, Eleanor, was there. I do remember that they had five children — the same as me, less one, and I was twenty years younger. No, it was Sandy, Hunter's wife, who was there and she was a worse singer than me — but not as bad as Hunter.

God, we were innocent. So was George McGovern, but we were in Washington to cover the Watergate hearings and nothing mattered except the moment. Anna had already spent three days around the pool doing needlepoint. I had made a line drawing direct on to linen of Senator Sam Ervin, head of the Select Committee to Investigate Campaign Practises, and Anna was unnervingly transforming it into a full-blooded caricature of the real thing. Like Madame Defarge from *A Tale of Two Cities*, 'waiting for something to fall into her lap', she sewed on and on over a period of ten days, creating the first, and probably the only, satirical embroidery in existence. *Rolling Stone* offered to publish it, as I felt sure they must. This was, after all, revolution — so why not needlepoint?

At that time everyone under thirty-five was hungry for hell. Now we have gorged on it, thirty and more years on, there is a strange nostalgia in the air, almost an apology. Nobody realized then that lifestyles have to be paid for, sooner or later. Well, maybe we did, but, looking forward from where we were in 1973, 'now' was way over the hill somewhere and faraway. A never-never land — and back then was sheer bliss.

McGovern was a warm companion and entered into the evening like a man used to fitting into any situation. Since he was committed to amnesty for Vietnam refuseniks and had been opposed to the Vietnam War since 1963, 'Masters of War' must have warmed his heart, even though we crucified it right there and then, over dinner. He had failed to make it to the White House but he may have felt that even among that orgy of drunken singing somebody was carrying the torch forward for someone else, capable enough maybe, but as yet unknown. The admirable thing about McGovern was that he nearly made a dream come true. He made you believe that you could actually turn on a TV set and believe in your President again.

George McGovern was the man America should have voted for in 1972. Perversely, Richard Nixon got a twenty million-plus landslide, instead, so what, in their darkest hearts, do people really want? What do they really think when they are alone in a polling booth? Their secret desires exert a terrible pressure and move the hand to vote for what they most protest against. People must have been shocked and horrified by subsequent events. They certainly were then as they watched the events of Watergate unfold like an old Band-Aid being peeled back to reveal the wound as it turned septic before their eyes.

Americans still vote with a crude hand, except perhaps when memories are fresh and they go off at a tangent and vote for Jimmy Carter, who thought being President was a fireside chat with a bunch of friends. To the nation's credit he is still held in some affection but he is dismissed as being too honest. Maybe he was the antidote to Watergate, an appeasement to shame, a vote in the hope of a clear blue conscience.

I had first visited as a lost, penniless artist in April 1970 and when Watergate finally happened three years later it confirmed what I knew intuitively. That somewhere beneath that squeaky-clean exterior there was something fundamentally rotten at work and a lot of Americans felt the same. At the time of the Republican Convention in Miami in 1972, I wrote my feeling across the skyline of a drawing. 'AMERICANS DON'T WANT THE CLEANEST WASH – THEY JUST WANT THE WHITEST.' The 'screaming lifestyle', as I dubbed it, disturbed me to the core.

A pathological fear of communism was driving America, led by Richard Nixon's government, to repress any sign of it with whatever measures it had at its disposal. Covert operations became commonplace. Watergate was merely the symptom of the disease affecting the body politic. American foreign policy was highly volatile, particularly in South America, where millions of dollars had been pumped in to oppress President Salvador Allende's left-wing aspirations in Chile, subsequently causing his overthrow and death.

And this at a time when America was trying, through that summer of '73, to purge itself of all covert activity, now that the game was up. The fact that it managed at all says a lot for its written constitution but not much for those who manipulated its declarations. Anyway, that was then and this is now.

There we were sitting around the pool of the Washington Hilton, watching a tiny black and white TV screen, a nifty, new, little model that Hunter had bought 'to keep a handle on the Watergate hearings at all cost'. He was busy haggling with an electronics salesman he had summoned to the poolside. The man was definitely confident of a sale. He had arrived in a slick suit carrying a large suitcase full of gadgets designed to make life bearable: bugging devices, miniature tape recorders, directional mikes and a small briefcase complete with portable phone, the purchase of which Hunter was seriously trying to negotiate and put on hotel expenses. 'Can you imagine having a little fucker like that, and you are lost in the desert? It could get you out of some serious trouble. It is a vital piece of equipment for people like us, Ralph – and it's only two thousand dollars!'

I don't think he ever got to use it in the desert or anywhere else, for that matter, since he was forced to return it, as an irresponsible and excessive extravagance. Between deals, he was eyeing the screen and filling in forms concerning an accident he had been involved in two weeks earlier while on serious journalistic business. His rented car had hit a Cadillac, broadside, at four in the morning. In those days he was a tireless round-the-clock worker; no lead was too small to follow up, drunk or sober. He had been at McGovern's house until the early hours discussing what had gone wrong with the campaign and his mind must have been preoccupied, for I have never known him to make a mistake in an automobile. Driving for him was an art. He must have been drunk because he went on about dragging Chuck Colson, a top Nixon aide, down Pennsylvania Avenue by the balls behind a huge car, to teach him a lesson. Only a drunken conversation would produce an

"IF YOU'VE GOT
'EM BY THE BALLS
THEIR HEARTS AND
MINDS WILL
FOLLOW"(CHARLES
COLSON)
Ralph STEADman

idea like that, because generally Hunter is not a violent man.
I believe it was provoked by a favourite quote of Colson's about
the American people. It declared: 'Once you have them by the
balls, their hearts and minds will follow.' Such was the level
to which American politics had sunk.

A lot of the hearings were watched on the tiny TV screen
around the pool, but once or twice we ventured forth as press-
men with a mission. I am afraid that I am intimidated by
formal institutions, owing to some unfortunate childhood
experiences concerning a sadistic Welsh headmaster with an
open brief on corporal punishment. Normally I don't drink
unless I need to, which is often, in this world, to soften the
dreams of reason. On this occasion, however, I felt compelled
to take with me into the Watergate hearing room a six-pack
of beers and a hip flask of Glenfiddich whisky. It gave me
the Dutch balls I needed to cope with the demons of official
administration, security and the particularly vicious level of
surveillance in evidence everywhere, enclosing this room of
revelations that was being watched by the whole world. I was
far from home. I felt sick and out of my depth but I was here
to do a job for *Rolling Stone*. Nonetheless, I was about to fuck
up and make a spectacle of myself.

It is at times like this that I am at my most cunning. I become calm and adopt a very English demeanour. I look into the eyes of the security guard and the innocence in *my* eyes withers the paranoia in *his*. But to operate this trick successfully and wave a six-pack in his face you must be innocent, as I, essentially, am and security guards, usually, are not. Which is why they get the job.

If we could ever hold a court where the underdog puts the powers of authority on trial, most of us would be sickened by the crimes that are committed in our name. That was what we were witnessing in that Watergate hearing, except that it was still authority trying authority. Thus, a certain amount of institutional pussyfooting was in evidence. A respectful cross-examination was fastidiously observed, in spite of the awful revelations that were emerging daily. I interpreted the quiet restraint as an American attempt at dignity, although I felt certain that from time to time any one of Sam Ervin's committee, the magnificent seven, wanted to scream out loud: 'You lying cheating bastard! How can you possibly show us all up like this?' It was, to my stranger's eyes, making all of them guilty by association. They were in the same business, after all, and most of them, the judges and the judged, were either

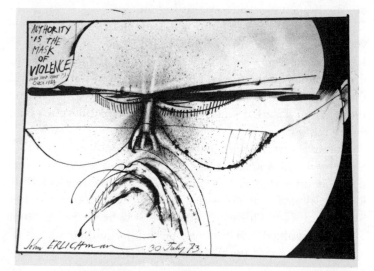

lawyers or accountants connected with the legal system in some way. They had all decided to go into politics and some had been perverted, or were already perverse. But where do you draw the line?

Those thoughts were passing through my mind as I watched the likes of Ehrlichman's and Haldeman's whispered conversations with their defending lawyers. 'Authority is the mask of violence.' Who said that? I don't know but it occurred to me and I wrote it down. I felt that the whole damn process was a charade and even if they got Nixon, some would get away with it and still be in place afterwards, inside the fabric of a system in limbo. Business as usual. I was thinking, too, that if they took this self-examination to its logical conclusion, even the Committee might have to testify and end up perjuring itself to safety or go to jail. It's a fine line and there were some things nobody wanted to know. Like the man I shared a taxi with on the way to the hearings, who said: 'It's like finding out your wife is running around but you don't wanna hear about it,' or as Melvin Laird, White House Chief of Staff, was reported to have said: 'If the President turns out to be guilty, I don't want to hear about it.' Watergate was a classic case of Hobson's choice. Whichever way you chose it the conclusion was going to be rotten. They were busy sawing off the branch everyone was sitting on.

By this time I was beginning to feel the rush of confidence you get when you chase each mouthful of beer with a slug of whiskey. I rose to move around and get near the rostrum with my sketchbook in hand as a foil, but with an idea in my mind that if I could grab a microphone I could make a few of my thoughts public right where it mattered – in full view of an entire nation.

I felt a certain righteous indignation and believed that anything I said would be fully justified since the object of this trial was, after all, that the truth will out. I guess I was a little drunker than I realized and I had not noticed that Senator Sam Ervin was on his way to the toilet. As he walked in front of me, I stumbled into him and he, being a gentleman, naturally assumed I was merely clumsy. There were TV monitor

wires stretching everywhere across the floor and it was no place for a drunk, particularly one with a beer in his hand. It was he who apologized to me for the beer on his coat and on the equipment that was strewn around like discarded possessions after a burglary, but it was I who realized, right then and there, that anything I said in that hearing room would fall on the deaf ears of everybody involved because everybody was involved, including the press. They were, to a man, more concerned that whatever the outcome, whatever the price, nobody could afford it.

I remember only that I was in a state of outrage because nobody would come clean. The domino factor was having a catatonic effect on everybody. The trial was a futile protestation against itself and its own wretchedness. Hunter claims that I had to be forcibly evicted from the room because of my embarrassing inability to play the game. I don't remember that, but if I was, then it would have been because I was the only person present who had nothing to lose by speaking out. In that respect I was dangerous. The proceedings would have been compromised, to say the least. The prudent course of action was to have me treated like some foreign stumblebum who made demented scribblings for no apparent reason and should therefore not be taken seriously.

The diagnosis of the trial, as we all know, was that cosmetic surgery was absolutely necessary. The President and his immediate aides should be amputated. The prognosis was that the body politic would recover in time and should return to normal. Whether the cancer was benign or malignant is still a matter of grave concern.

Twenty years on, however, the patient is still alive and at present is on a life support system – like the rest of us.

SAN FRANCISCO, 14 AUGUST

In her diary, Anna wrote this record of the next two weeks:

> I won't try to write everything down in its correct order because so many things have happened.
>
> We have been back in San Francisco for nearly ten days. Ralph has drawn twelve or thirteen Watergate pictures. At first he had to struggle to get into it but now he is nearly on the last one. He started with *The Lie Detector* – Haldeman's graph shows a picture of Nixon. He did the Colson one that Hunter wanted him to do – 'Grab them by the balls and their hearts and minds will follow'.

The Caucus Room drawings are very strong – the snakes testifying, Patrick Grey twisting in the wind, lawyer Wilson advising his client, vultures of the White House – also Ehrlichman's face and *Nixon Crucified in a Stained Glass Window*. Jann wanted one or two colour pictures but Ralph had decided to limit the use of colour as he saw fit.

I think he sees Watergate in terms of black and white television images. Last night he stayed up till two doing a complementary drawing to the one of the snakes testifying in front of the Watergate Committee. It shows people in a bar watching Watergate. The germ of this idea came from Dan Ellsberg who said that Ralph had missed out one thing – the happiness with which people turned on their TV sets to watch Watergate, or read about it in the newspapers. It was a picture of boredom on the faces of the American family watching TV.

Having returned from Washington with my embroidery of Sam Erwin nearly finished, Jann seemed very keen to use it for the *Rolling Stone* cover for the issue about 'The Wit of Sam Erwin'. Annie Leibovitz had it photographed and all seemed fine until nothing was said for a couple of days. Jann had a guest art director called Dougall Stirmer who was with *Ramparts*, an early, left-wing, political magazine, similar to *Scanlan's* and *Evergreen* and he did not use the embroidery. However, I did get a cheque for 450 dollars which is rather nice, I must say.

For three days there was a very distinguished visitor in the house, Daniel Ellsberg, who took and made public in *The New York Times* the Pentagon papers, which gave the US military's account of activities during the Vietnam War. For those three days, Janie, Ralph and I had to walk on tiptoe while Jann and Ellsberg were tape-recording a three-day interview. Ellsberg seemed to be miles away and never seemed to know who was who. A friend of the

Wenners called Ned happened to arrive as Dan was leaving. Jann introduced them on the front steps and Dan shook his hand and said: 'Congratulations', while he looked blankly at Janie when she said politely: 'Please come again.' He met me in the upstairs corridor on the second day and said: 'Have we met? I'm Dan Ellsberg.' I think he thought I was another maid. The thing that made us laugh was that he was always sneaking about and we joked that perhaps that's how he managed to get to the Pentagon Papers. He also used to leave his bedroom door open with his books deliberately spread over the floor as if to show anyone passing just how widely read he was.

On Saturday we drove out of San Francisco with Janie, Jann and friends, Ned and Kathy, to the mock-Elizabethan manor house that Kathy's parents have just finished building. We swam in the pool and sat in the sun all afternoon. The next day we were invited again. This time with Art Garfunkel. He came with his wife Linda at about midday. Janie had told us beforehand that she had the feeling that

SUMMER 71: PENTAGON PAPERS PUBLISHED IN N.Y. TIMES. NIXON WOKEN FOR SPLIT-SECOND BY TRUSTED AIDE JOHN ERLICHMANN (FORMER 'SEVEN-UP, SANI-FLUSH SALESMAN) NIXON HAS COUGHING FIT AND RELAPSE.......

Linda didn't like her. Linda was very quiet and uncommunicative. When we all got back into the car to go to the pool, we stopped outside a hotel and Linda got out. She didn't say goodbye or anything.

ASPEN

After a couple of weeks of nationwide sightseeing, the last stop on our trip was a visit to Aspen. It was time to see Hunter in his own environment. How did he behave at home? How did he cut it with the locals and how decent was he to his wife and son?

My observations are very personal and Hunter did not go out of his way to hide his nasty habits. I suppose it is best to discover one's friends' proclivities as soon as possible to ensure that they are habits that can be forgiven at a moment's notice. All of his habits were forgiven by me almost instantaneously. And they were many.

For instance, he rarely wore socks; so his feet stank. I don't mind that, because most people's feet stink, off and on, if you hang around with them long enough. It is quite a natural condition, considering that we all eventually get shaped and

crafted by the job we do or the service we perform. Whores get bow-legged and bankers get mean, which is strange when you think that if whores get bow-legged, bankers should get generous, but they never do. They always say that it would be more than their job's worth. So, I reckon most whores are generous. In defence of Hunter and his Converse Low trainers, many times I watched them spin, clonk and bounce in his washing machine in the cellar.

He had a mynah bird called Edward. This bird could speak and its favourite phrases were: 'What's goin' on?' and 'It's . . . I . . . er . . . not sure about that,' but not much else. You would think that maybe a wordsmith of Hunter's calibre could have imbued the bird with a spot of Mark Twain, Walt Whitman or maybe even a line or two from the American Constitution. Even 'We the People' would have given dignity to the bird, but nothing ever left his beak except those phrases and 'Aark!' and 'Squawk!' Maybe Hunter could have taught it to say: 'America always! Always our own *feuillage*! Always Florida's green peninsula! Always the priceless delta of Louisiana! Always the cotton-fields of Alabama and Texas! Always Owl Farm, Edward! Always Owl Farm! Say Owl Farm, Edward! Say Owl Farm. Fuck you Edward! Say it!' But usually he only said two sentences and it only happened when Hunter advanced on the iron cage and opened it with menace, whispering dark threats and veiled prophecies of doom.

I remember one particularly intense demonstration — it was a 'demonstration' because it was staged entirely for my benefit, and was done with a showman's style. Hunter reached into the cage and grabbed Edward savagely, dragging him out and growling: 'Edward! . . . as far as I know, Edward! . . . there is not a bird-God who is going to save you now, Edward! No, Edward! You are doomed, Edward! You are doomed!' The poor bird fought back, demonstrating his own form of self-preservation. He tore and pecked like a maniac and kept his beak in a particularly threatening pose waiting for the next barrage of menace from his master. 'I'll get you, Edward. You mark my words, Edward. You are doomed!' Then he would release the poor creature back inside his cage, go to the fridge and get himself another drink.

Edward had been chosen — like any pet — impulsively and from what was available at the time. Like any pet, except for Hunter's cat Jones . . .

Jones chose Hunter, or at least, Jones chose Hunter's house. It suited his purpose right down to the tip of his bushy tail. So Jones stayed and, by the looks of things when I showed up, very much on his own terms. He and his master shared a deep and guarded respect, a mutual recognition of each other's stealth in matters of survival.

I have never met a cat more insolent and yet more appealing than Jones. He never needed to judge a situation. It took me two days to feel the full weight of his authority, and when I did, I decided to buy a handsome sketchbook from a store in Aspen. I needed to capture something of him, and the best way I know is with a few direct lines, straight from the eye through the mind to the hand. The result on paper can be fiendishly perceptive or hopelessly inaccurate, but it is always an intriguing and playful possibility. In my drawings, Jones decided the style and I went

along with it, being the weaker of the two parties. Jones had a manner that engaged many people. His remoteness was a spur rather than a hindrance to natural affinity. He found his place and knew it. He coveted nothing but his own comfort, and lacked only a god's control over his own fate.

In those two weeks, odd friends came and went, stayed and dabbled, made arrangements, laughed and ate, slept on sofas. The actor John Belushi tumbled through on one weekend, reeking of fatigue and rocket-fuel adrenaline. He buried his face in cushions as if burrowing in search of relief from his bruised brilliance. Frenetic outbursts burned him up like a Roman candle. Jones treated him and the rest with similar détente.

Then I, like all the others, left. For all Jones cared, we might never have met. I was just another passing entertainment.

Later, Laila (Hunter's girlfriend at the time) came to visit me in England. 'How's Jones?' I asked.

'Jones is dead,' she told me. 'And that's how I heard it. Just like that.' She had phoned Hunter, and asked, as I did, 'How's Jones?'

He struggled to tell her gently, but hesitant explanations triggered the desire to get it over with, and the truth tumbled out abruptly. 'He's dead. JONES IS DEAD!'

Laila told me how he died, but I forget and maybe I don't really want to know. I wanted to ask Hunter the next time I saw him . . . Now, of course, I never will.

Hunter 'demonstrated' many times. He would grab the ear of Juan, who may have expressed an individual view of his own, and swing him around the room like an average-sized cat. Juan watched his father's every move, avoided some of them and respected the nature of his given relationship with a man who was not seriously going to play father in the conventional way. Some would say that Juan was lucky to have such a colourful father and Juan secretly appreciated that unchosen role, and I saw nothing uncommonly vicious. There was a secret tenderness that I believe Hunter and Juan enjoyed as though the outward signs of distance and malevolent behaviour were put on strictly for visitors.

We decided it was time Juan had a day out. We would go fishing and have a picnic. Sandy gave us her full support. 'Hunter never takes Juan anywhere,' she said. 'Juan would love that.'

'Yup!' said Juan.

'You like fishing, Ralph?' said Hunter.

'Some people fish,' I said, 'but they catch nothing. Others fish and the world is their oyster. It's all a matter of philosophy. One day you will learn. One day you will pick up on my finer nuances of life's chance.'

'I believe you this time, Ralph, but your time of disappointment will come. Good luck!'

We went to the superstore in Aspen and bought a couple of pounds of sausages, some cheese, bread rolls, salami and fruit. Juan was ecstatic and loved the idea of preparing for this as if it was a special event of great importance. I don't think his father ever took him anywhere but here we were, and he was very excited.

There is a river in Aspen called Roaring Fork. Not a huge river, but fast-flowing with deep pools at odd places and access

to some of the more promising riverbanks. We found a perfect spot where we could cast our rods and more importantly we had a flat piece of ground where we could make a fire to have a sausage picnic. Anna was content to sit in the sun like a goddess and watch. This was her special treat. The boys were content to cast lines, spinners with worms to add an enticement to the fish. It became obvious after the first hour that we weren't going to catch much. You get a sense when things are going wrong. What the hell! – we were having a good time trying and we bonded like good buddies.

'Never mind the fishing! Let's have a picnic. We need some wood.' There was plenty of that and we built up sticks around the pages of *The Aspen Times* I had bought in the store. This was the fun of it all. 'We'd have let the fish go anyway,' said Juan. 'Otherwise we would have had to kill them, Uncle Ralph!' I don't think Juan would have enjoyed that part of it.

So we cooked our sausages and had our cheese, bread and salami. We had a lovely meal, but there was one sausage left. This is the moment when I saw the sense of mischief that Juan had inherited from his father. Instead of throwing the sausage into the undergrowth, he said: 'I've got an idea!'

He found a flat piece of wood, flat like a piece of orange box. Then he took a length of fishing line, and carefully bound the sausage to it, which gave weight and balance to the wood. Ensuring that the sausage was firmly attached to the piece of wood, he waded out about three yards, placed his raft on the water and then let it go. 'Hmmm . . .' he pondered as he waded ashore. 'I wonder what they'll make of that downstream?' It was a wonderful surreal thing to do and it made my day.

We had a lot of fun for the rest of our stay in Aspen and my work for the Watergate hearings was done. It was time to go home.

RUMBLE IN THE JUNGLE: ALI v FOREMAN

30 October 1974

No room at the Intercontinental, Kinshasa . . . A
midnight chat with George Foreman and Joe Frazier
steals my pen . . . Hunter sells our tickets and jumps in
the deep end

In my studio, after my trip to the Ali–Foreman fight in Zaire, I wrote: 'Now I must write.'

I would scribble stuff onto large pieces of drawing paper and send them, with the drawings, to the editorial department of *Rolling Stone*. Hunter hated it when I tried to write anything. So it was a smuggling activity. If it had anything to do with one of our joint efforts, it never worked. Hunter was worshipped and his words were treated like the Ten Commandments and he was Moses. He would say: 'Ralph can't write!' which was true. But curiosity at least compels one to try. I always said that I could write better than he could draw, which was also true. But when it came to writing in any form, Hunter got mean and possessive and he had every right to be, screw him! There was no chink of light for me anywhere. If Hunter had wanted to draw, which he did, and take photographs, which, incidentally, he was very good at, why didn't he? I would have welcomed the opportunity for him to try, just so that I could mock him. He mocked me horrendously and mockery was an entirely new dimension for him after our first piece together on the Kentucky Derby. Here was this innocent from England whose daft comments had a certain edge to them but only because he was too steeped in his American Way and, like millions of ordinary as well as some thoughtful Americans, had forgotten what lay beyond the perimeter of their fractal shores. They had forgotten that the rest of the world lay there. That is exactly why America has no sane foreign policy to this very day – why the French heaved a huge sigh of relief when the US took on North Vietnam and found themselves in the biggest mire of jungle warfare they had ever known. They sure as hell didn't know where Vietnam was, until thousands of body bags, containing the remains of 'their boys', started to arrive back home in disquieting numbers. Nobody knew where Vietnam was and neither did they give a shit – all they knew was that their boys were dead. It was the great awakening for America and the beginning of a protest movement that is still alive today.

Try to imagine that there existed at the time millions of immigrants who had left Asia Minor, Europe, Ireland, even England, to find a life in the 'New World' and were suddenly confronted with a vague involvement in something they thought they had left behind. This was their biggest nightmare and yet thirty years later they were suddenly confronted with a new menace. A black man, Muhammad Ali – a dissenter – was going to fight another black man, George Foreman, in the blackest place on earth. 'We must make it fun,' they thought, 'a noble event, for the sake of the soul of America.' Now they were on to a winner – two black guys hating each other enough to beat the living shit out of each other, prepared to go back to their homeland and no matter who won it would be a triumph for America.

Black men and black women were 'going home' to show the world that poverty was history and there was not even the whisper of a colour bar. This was post-Watergate, when every white bastard who had clawed a position to the top of the heap appeared to retain the great democratic vision of equality. They had made it and hung on to their vicarious vision like limpet mines about to explode and destroy the forces of evil. I wish it were so, if it were true, but like everything else, it was merely bolstering up the heinous soul of a dirty, low-down, profit market, watching every punch like a blow struck for American supremacy.

My initial effort when we went to Zaire for the Ali–Foreman 'Rumble in the Jungle', began as a handwritten diary. It read:

I caught a particularly virulent hot dog in Bruxelles airport. I called the airport authorities and, they shot it.

Announcer at Bruxelles Airport whispers off words of flight numbers to you like she's on the next pillow.

Suddenly at Gate 6 (Bruxelles) my fellow passengers are lay preachers, missionaries, black professors, moustachioed 'Old Bills' and strange, uncertain no-nos of extremely well-fixed addresses.

As I board the plane I breathe a last deep breath of fresh air and enter a stuffier clime – disagreeable body odour lays heavy on my clothes and I don't think it is me (yes, it is as well). Thank God for the refresher towel given as a hand-out.

Where the fuck are we? On the way to blackness, I think!

Bugs everywhere. Wherever I sit, so does a bug. We are now in Nigeria – which is part of Africa. Need a beer. Transit Bar – who needs Nigerian money?

The barman in the Transit Bar accepts a pound and gives one-arm bandit discs in exchange. OK, so it's £1 for a beer. Let's get on.

5.30am. October 74. Arrive.

Oranges in the trees. Gentle ride from airport to Intercontinental Hotel along straight, flat plains.

Africans – girls with baskets on heads saunter along – people wait for hundreds of yards along the mud sidewalk for something – intermittent – but continuous – like an irregular pulse – I don't know why they wait and I couldn't ask.

Dull dawn sky – slight drizzle. Taxi is broken down but it works somehow.

Arrive at the Intercontinental Hotel telling the

Welcome to Kinshasa 05.30 hrs. 24.10.74. EXIT

Ralph STEADman

driver in crippled French that if he waits I'll change
a Bank of America cheque and pay him.

No room recorded for me. Somebody has fucked
up again. I argue a little – gentle and beyond reaction
I wait.

I wait.

'Ralph! Goddamit – where you bin?'

Hunter, whisky with ice in one hand, credit card
wallet in the other, sweating, early morning red eyes
– weighs in heavy – my natural, gentle English
manner restrains him from pulling the head off the
concierge, avoiding a complete showdown.

'We can't change your travellers cheques, M'sieur
(*nous ne pouvons pas changer vos cheques, M'sieur* – not
bad for a cripple). You don't have a room here (*nous
n'avons pas une chambre ici pour vous*).' '*Mais oui!* I do!
Rolling Stone – half million sales in America alone –
booked me in – assured me – guaranteed it.'

'Sorry sir . . . please . . . 10 minutes.' I just hit the
strange lethargic NO that stops your adrenaline dead.
You can stand in the same mind-given length of time
amidst any conversation, or any human twitch you just

had and look at each other — him on one side of the counter — and me on the other — immovable. Irresolute! And at 6am on a Sunday morning nobody moves from their appointed spot. Not him and not me now!

Hunter moves, but he is already deep into something else; I have no possible inkling of what. I go along.

I follow him out of the lobby, carrying everything with me and stagger into an Avis car. Carrying all my luggage, the rain covering me like the fear that at this moment in time is taking hold of my mumbled brain.

Already, this is it? Not already! I'm innocent. I've just arrived.

Hunter has a friend with him — Bill Cardoso of the *New York Times*, masquerading as a *R. Stone* reporter. Stoned after six weeks of Zaire, heavy dope and no way home. Happy — sort of — but resigned and a tiny bit hopeless.

Hunter starts the car in gear, stalls, starts again, grinds on first — no — reverse — backs up — jerks forward — drink in hand (left) spilling — credit cards flutter everywhere. Bill sways and snuggles up to Hunter involuntarily as the car tears away and out of the car lot. Bill sways again — Hunter misses a turn — swears, spills more drink. Does he really know what he's doing?

The car is in second all the way — engine revving — hot oily smell. Car is on fire — or should be.

6 blocks one way, 3 another, half a turn this way, 2 double backs, 4 wrong turns.

We arrive in flames. The club is on the corner (remember, it's 6am).

Bill disappears. We look for him in a dirty early morning room. Just the smoky dreariness of afterwards.

No one but a man finishing the dregs of every glass in the place and us looking about — lost and up for grabs.

'Where's Bill?' Hunter asks. Why Bill — we just had Bill in the car — What we doing here? Look for Bill.

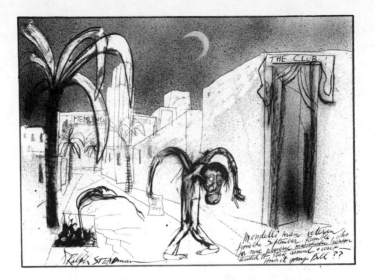

Why? Who's Bill? See him — two blocks away — hunched — droopy — unsteady — walking home in the wrong direction to the Memling Hotel.

Follow him — how many blocks? — try three — we practically run him down.

'Hi! Hi, man! I'm just going home — gotta sleep.'

'Steadman needs a room!'

'Be my guest!'

'Just one moment!' say I. 'Where are we, why and who? — OK never mind. Who needs it anyway?'

Hunter by this time is past anything sensible. He pulls out a container of Nivea skin cream — dips his fingers deeply into the white glutinous glob and smears it down his throat like a frog swallowing a fly. 'Got a sore throat,' he said. 'This filthy African humidity — like sucking on a swamp.'

Hunter and I leave Bill and drive back to the Intercontinental. Drab newness and drizzle in the turgid heat. Hunter has already read my predicament — my pathetic posture. OK. I have arrived and just for a moment Hunter takes control and has words with the Receptionist. 'No Room!!!? What kind of a hotel is this??? We don't just arrive from half way around the world and then fuck off!!! What is the tenure on my room?'

'You are booked in for the week, and it is a double room – m'sieur.'

'Hot damn, Ralph! You can sleep in my room. Do you have any weird habits?'

'No weirder than yours, I suspect.'

'Have his bags taken up to Room 824, and, er – you have your American Express card with you?'

I did, and all that was settled.

'Let's have breakfast. They serve whisky and Daiquiris when you have the house snake omelette. Just kidding, Ralph!'

'Mine's the snake omelette,' I said. 'I always have that when I'm tired – and I need to change some travellers cheques and my underpants.'

'No time for that now. Let's have breakfast and then we can get to Ali's training camp. I need to ask him a few questions.'

Such was the security pulsating through every jungle clearing in Zaire that we never got to see the training camp of Muhammad Ali or any one of their party. However, in the dead of night, at four in the morning, a party of George Foreman's protection posse paced George through the corridors on the eighth floor of the Intercontinental Hotel, Kinshasa. At that point no one dared to interrupt the champ's routine with something as banal as an interview or even a few enquiries delivered while running along beside him, but I fell in step with him just to ask a simple question. The posse moved forward to stop me but realized from the beatific smile on my face that I was clean. I wanted to ask the simple question: 'Whotcha gonna do when this is all over and you are the champ and . . . er . . . you'll have a lot of time on your hands? Nobody fights forever.'

'Yo're right man. Yo' the foist to ask me dat. I'm goin' inta da dough business.'

'The dough business? Aren't you into that already? Five million a head right here . . . win or lose?'

'No, man, the *dough* business. I'm gonna make bread like my momma used to make wid her bare hands. It's sommink I can do. I can beat the shit outa dough. I'm gonna be a baker.' In fact, a grill machine was famously named after him years later and he advertised it on TV, making a lot of dough from it.

'Oh, that! Of course. What a noble ambition, and I really mean that, because all this schmahasagas will pass − you have it all worked out. Good man! Good luck!' I couldn't keep up and slumped against the door of room 824.

'Who's there? What do you want?'

'It's me! Ralph! Let me in. I've got a scoop!'

'Only hack journalists get scoops, Ralph! What did you get?'

'Only a one-to-one chat with George Foreman. He wants to be a baker.'

'A baker! What kind of insane bullshit is this? You spoke with George Foreman about bakery? Okay, Ralph, I knew you were perverse but this is insane.'

'Honest!' I pleaded: 'It's what he really wants to do. He can't be a boxer for ever.'

'You're right, Ralph, as always. But what we need is the real story behind this scam. Where did the money come from?

Why does John Daly have such a low a profile, or, indeed, is he even in the country? I heard he has been arrested. How does David Frost fit into all this? He's not exactly known for his love of boxing.'

John Daly was one of the shadowy businessmen behind the fight. 'He's a satirical businessman,' I said. 'He does everything for laughs, which makes sense. If I had his money I would want to squander it all on an impossible dream and all this seems like Mobutu's impossible dream – to put Zaire at the forefront of world politics and sporting prowess.'

The next day I took a leisurely, observant walk around all the bars and coffee dens scattered throughout the hotel. Most were full of journalists with theories, but the bars were surprisingly empty. It has to be remembered that the original fight had been scheduled to take place six weeks earlier. Journalists had poured in from every corner of the planet and the intensity of public interest was at breaking point. I don't think many of the journalists had been to Africa before and Zaire's President Mobutu was intent on milking everyone for everything they had, including their newly acquired currency, the zaire. Conspiracies were everywhere.

Then, during one of Foreman's training bouts, he sustained a cut from a sparring partner. It was an accident, nothing suspicious, but the rumours ran wild and accusations followed and grew into gospel truths that the injury was a fix to give Ali more time to train. A majority of the journalists went home, intending to come back when the fight was rescheduled. Some, however, stayed on and succumbed to the easy lifestyle of drugs, cheap beer and sex. We had waited at home, however, until the fight was rescheduled, but we expressed our doubts.

On 1 October 1974, I had received a hand-written letter from Hunter – a wild and angry broadside to my suggestion that we should perhaps cut our losses and leave the fight alone. It was hardly rock'n'roll:

```
'. . . have you lost your fucking mind? We
have to be in Zaire on Oct. 30 for the fight
```

- we've already paid $1000 deposit for our
hotel room. I'm keeping the same plane and
hotel reservations I had - but changing the
dates to Oct 23 thru Nov 1. You should do
the same - tell Bailey [Andrew Bailey – English
Rolling Stone editor at the time] to fix things for
you - and also to update my arrival in London
to Nov 2-15 - same hotel, same typewriter,
etc. I promise your drawings will fare better
this time - if they're any good, heh! You can't
quit now Ralph - we have to go thru with it.
Call me here in W.C./OK - H

I had moved into Hunter's room and several times a night
there would be a knock on the door. Answered always by
Hunter, a voice on the threshold would ask: 'Can I have some
more medicine, man. I really need some medicine.' Hunter
had purchased a huge plastic bag full of African grass, for
the equivalent of about forty dollars. It was about the size
of a regular golf bag and there was plenty for all who came
in search of 'the medicine'. Some had been in Africa since
the July cancellation, had got into the grass and anything
else that was going and now wandered around with faces
like African voodoo masks, looking for the next fix. Serious
junkies after six weeks is quite impressive. Hunter had shown
great foresight, which helped his own profile. Many had lost
any memory of why they were there in the first place, but
had always managed to wire back to their papers for more
funds to enable them to get the story of a lifetime. Some
had gone native. The stories were all dope-fed
fantasies.

Wandering into a bar one morning, I ordered a beer and
whisky chaser which was my tipple at the time and moved to
a corner to watch for scenic interest. Sitting at the end of the
bar was Ali's cheer-leader, Bundini Brown, an old childhood
buddy of the former champ, or so the story goes, whose job
was to keep the circus on the road. His real job was to berate
and demoralize anybody who chose to come to the bar for a

drink. After an order had been placed, Bundini launched into his tirade, which I wrote down verbatim:

'Yo' put yo' money where yo'ass is, Whitey. Blue eyes, green eyes, sees green grass – yo' motherfucker yo.' Yo' kiss my back – I'm yo' daddy, motherfucker. Yo' my son! I'm gonna kill yo' if yo' say I ain't yo' daddy – fuck yo' motherfucker – we gonna let yo' kill yo'self. Yo' is fucked. We from the root to the fruit.' Many moved away to find a small corner to hide. I thought his outburst was pure Gonzo and a great subject for a drawing. I went back to the bar and said: 'Same again, barman!' Bundini got right into the same spiel like a record and went for me. 'Have one on me and thanks for the idea,' I said and stuck it out until he really did begin to sound like a stuck record.

Hunter wanted to go to the market for something unusual. We were crowded by people trying to sell us anything and everything.

'I want something I can't get in Colorado.'

'You can get most things here that you can't get in Colorado.' Good point.

'How about an elephant?' I said.

'Good point, Ralph. What I need are some tusks!'

'Yeah! Good thinking, Hunter,' but he was off looking

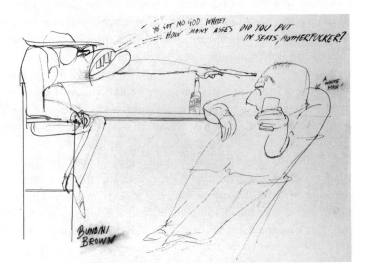

around, like a rat after filth. He found a tusk-seller — it was still legal to sell them then — and wanted to buy a pair he had on his stall. There was a lot of banter and three hundred dollars was mentioned. A deal was struck, there and then, and out came a wad of non-negotiable American Express traveller's cheques. The poor bastard trying to strike the deal just saw the dollar sign and was transfixed. This was his lucky day! Hunter showed his passport, assured the poor man that this was indeed his lucky day, and signed off a worthless token in exchange for the tusks which, as far as Hunter was concerned, were what he had come to Africa for in the first place. (I have a poor Polaroid of the transaction actually taking place.)

We saw the week through, and spent hours by the pool, watching TV footage of training activities, interviews with Don King, the promoter, the trainers, films of bouts and Ali giving some of his great monologues. There was one memorable moment when one afternoon a light aircraft flew over us pulling a banner which bore the words: 'MOBUTU WELCOMES YOU TO KINSHASA'.

'Oh, God!' said Hunter, a half-smile on his face, but held back, a faraway look in his eyes, as though he was thinking ahead. He was.

'I'd like to hire that fucker up there for an afternoon,' he said. 'Oh, yes, we could write a screamer for that ourselves, Ralph, with those big felt pens of yours. Something to send shock waves through the nerve ends of everyone here. Something like' — and he was looking down through the floor, as it were — 'something like "BLACK IS WEIRD".' Don't forget the feints and nods with the head as he said this — his head jerked back when he lit a cigarette, like he was short-sighted and couldn't see the tip of it.

The fight was scheduled for ten at night, so that the temperature would be marginally cooler, and by the time it came round, most people were in a frenzy of anticipation. The door to our room was busier than usual and the huge bag of grass was fast disappearing, which is to say that the bag was still half full, rather than half empty.

BLACK IS WEIRD

H.S.T. in the pool during the 8th round. 30.10.74

I couldn't figure out why someone wanted to be so stoned to watch the greatest confrontation in living memory. I decided to take a walk and wandered around in the hotel lobby, people-watching. Damn me, there was Joe Frazier! Impulsively I went up to him and asked for his autograph. I offered him my sketchbook and he held out his hand, until I thought: 'Oh, hell, yes − a pen!' He scribbled his name on the white space and handed me back the book but put the pen in his pocket. Do you ask for your pen back or let it ride? I was pissed off and only glad it wasn't my Parker I had lent him. Not the most pleasant of people and I just walked away, went to the bar and got myself a drink. There were a few huddled groups there and I guessed last-minute interviews were being conducted. Plenty of time to go back to the room before the fight. I had my 'scoop' interview anyway.

I must have had a few drinks because a crazy idea entered my head and I felt bold enough to try it. I went to a phone booth and got the operator.

'Number, please,' she said.

'Er, could you put me through to President Mobutu or one of his staff?'

'Who is this, please?'

'I am from the London *Times* and I need to speak with him.'

'I can't give out a number like that — even if I knew it — but I can put you through to his Press Office.'

'That'll do fine!' I said. What brought on this maniac notion I will never know. I certainly wasn't high but my idea was to get myself invited into his inner circle, watch the fight with him and get myself the first-hand reference I needed to do something spectacular. As I write this I am wondering if I did this out of sheer desperation or from a weird premonition that Hunter might have sold our tickets, two of the best ringside seats in the place — on the very night of the fight. It would be the dumbest Gonzo thing he had ever done. What I was doing was responding in the only way I knew how. I was generally so pissed off, anything was worth a try. A London *Times* artist sitting and watching the fight with the President himself? Why not? There were several interchanges with different voices and then one asking: 'Where are you?' It was like I had hit a nerve of interest.

'I'm in the lobby,' I replied. 'Can somebody pick me up?' I promise I will keep very quiet and simply observe.' I waited crazily by the front of the foyer, looking for what they had described as a dark green Jaguar — but what was I doing? What on earth would this achieve? I waited a considerable time and then something whispered in my inner ear: 'This is suicidal. Do a drawing instead.' I imagined the scene instead and wrote a caption for it.

Surely my most humiliating moment (even more than having to pretend whining like a lost corgi). Picture of me hanging upside down — strung up by gribbling roots, watching the whole 8 rounds, before 'they' cut me loose — revived me with a tea made from IPOMOEA (Morning Glory) and sending me stumbling into a sheet of dawn rain. Humiliation seems to fascinate Mobutu who watched my ordeal without once looking at the closed circuit TV installed for the fight.

When I got back to the room, Hunter was on his way down to the hotel pool. The fight was about to begin but he didn't

want to know. He had a towel and a bathing robe, a bucket of ice, a Heineken and a bottle of Glenfiddich, my present to him when I arrived. It became one of his favourite whiskies.

'What's going on?' I asked.

'Nothing, Ralph! The tickets are gone. Sold!'

'What?' I couldn't believe it. 'What are you going to do?'

'I'm going swimming, Ralph. Come and watch, unless you want to watch the fight on television, 'cos that's all there is now.'

He asked me to carry the bag of grass down to the pool, positioned his ice bucket and the Scotch and Heineken by the poolside, emptied the grass into the pool and dived straight into it. It floated on the surface and moved very slowly, but definitely, towards the pool filter and Hunter emerged through it.

'This is it, Ralph. An aesthetic experience. Fuck the fight! If you think I came all this way to watch a couple of niggers beat the shit out of each other in a rainstorm, well, you've got another think coming. The fight's not the story. Where did the money come from? That's our story, Ralph.' It was inspired madness and, come to think of it, I wasn't all that keen to see the fight — there would be replays, anyway, *ad nauseam* for weeks to come. I got it all from there, and talk of gun-running from Brazzaville just across the Zaire River added to the crazy romance of it.

I didn't care either. What I did care about was something weird emerging from all this talk of the Black Man coming home to his people. Patent leather Mendelli man (please don't look for this in the dictionary — I made it up), as African-Americans were called, trying to bond with their brothers in the middle of Africa. There was something daft about the whole event, but someone was making money out of it for sure and it sure as hell wasn't the ordinary folks! As soon as the fight ended, the huge exodus began. Get to the airport soon as you can, and leave behind every single coin in your pockets, foreign or otherwise, and get the hell out.

'Don't even think of hiding one single zaire, Ralph,' warned Hunter. 'We are all going to get body-searched and will end up in jail. Not one single zaire, Ralph. And don't even dream about hiding one in your nuts. That'll be the first place they'll look.'

So, there we all were, trying to get on the Exodus Special. Norman Mailer was there, George Plimpton and Hugh McIlvanney, to name-drop just a few, and every sportswriter from the western and the eastern and the southern and the northern hemispheres, eager to get airborne and away from a place that twelve hours earlier had been *the* place to be. President Mobutu had anticipated this. After such a spectacle his kingdom would wilt to nothing and with it all the sources of ready income would be sucked away in the slipstream of a flying cornucopia.

Everyone got the treatment, some more than others. Spare change clinked on the floor and no one but officials were picking it up. It surprised me that we got out with our credit cards, but even Mobutu reckoned that would be going too far and might even put him in the dock.

The plan was for Hunter to fly back to England with me to finish the story in London. We would have to stop over and get another flight from Lagos to London the next day, but Hunter started to grumble about even one night in Lagos. We

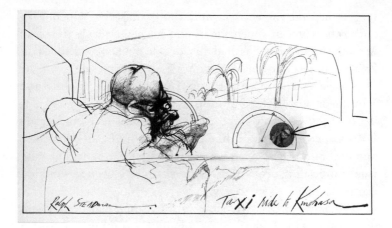

Taxi ride to Kinshasa

would end up in a filthy rat-infested jail, never to be heard of again. We would be sold into slavery.

'These people don't joke, Ralph! When we stop to get off, just grab your bag — and follow the pilot.'

'Follow the pilot?'

'Yes! I've got an ugly feeling. I think they've already impounded my tusks! Remember, just look down and walk with a purposeful air. Don't look suspicious. Put your shirt-tails into your pants and move or we'll never see daylight again. Don't look sloppy. We know what we are doing Ralph. We are going through security with the crew and out the other side and back onto the plane.'

'But I haven't got a visa for the US.'

'Fuck your visa! We can deal with that when we get to JFK. Stay here and a visa will be about as useful as a bent credit card.'

Tired and dishevelled as we were, we had moved to the front of the plane and watched for the crew to disembark to register whatever flight information was needed before take-off again for the US. I was amazed how easy it was, except when one of the stewards on the flight turned and wondered what we were doing following them.

'We're getting back on the flight,' insisted Hunter. 'This was a bad mistake. We have seats. Fuck it! Some are even sitting in the aisles. Desperation is rife. White people are easy targets around these parts and there must be someone in authority who can vouch for us and let us get back on board. Mr Steadman here has had his tickets fucked with back in Zaire. He needs to get to New York in a hurry.'

I have that steward to thank. He nodded uncertainly and mumbled: 'Just stick close to us.'

Back on board we were welcomed with jeers and crude jokes about being unwelcome pimps looking for new business and how our sort were not wanted in Lagos. Drink was flowing freely and safety precautions seemed to have been left behind on the tarmac.

When we arrived in New York, I then had a problem explaining why I had even thought of coming to the US without a visa. I would have to stay in transit until I could get a flight out back to England.

It was then that Hunter started bitching about his wretched tusks. They had made it to New York, but Customs had impounded them. He was raving and protesting as a wronged US citizen who paid his taxes and upheld the Constitution. Meanwhile he stayed with me, partly, I think, because the Customs storage department was right in the transit area and this was his big chance. He had actually seen them, just on the other side of the low tables where they check your baggage if you get stopped.

We found an open bar with no door, in one of the wide corridors that connected different gates and terminals. I still had my bags with me, to check in again. We settled at the bar.

'Okay, Ralph, this may be tricky but I'm going to try it. I'm going to get those tusks, but I need your help.'

'I'm really enjoying this beer, Hunter, and the whiskey is going down a treat.'

'Exactly, Ralph. Just stay like that, enjoy, and pretend nothing is happening.'

'What are you going to do?' I asked.

'Something unexpected. Something that will require nerve, calm and precise timing. Football has maybe been worthwhile after all. Look, you just sit here minding your own business. Forget me, Ralph, but be prepared for a swift and elegant movement.' He held up a stern warning finger. Well, 'warning' is a bit strong as a word. More of a 'be prepared' motion. 'For justice, Ralph. For justice.'

I sat nonchalantly at the bar, rolling another Golden Virginia cigarette. You could smoke anywhere at the time; they were blissful days and security was lax. It was a time when you could race to an airport anywhere in the world at the last minute, and they would hold up an aeroplane just waiting for you. Passengers were Lords of the Grail and were important. Today we are sacks of shit bundled into flying tubes with a security warning secreted inside every orifice.

Hunter loped off towards the Customs holding area, slowly at first and then gradually building up speed. It was a fine sight to see, an elegant figure moving in slow motion. The officers saw him coming but did not expect what happened next. Lots of people run through airports to reach their gate but, with a full frontal movement, Hunter took off, rising off the ground, over the tables, between the two officers on duty and through the wide-open door ahead. Already he had scooped up the two tusks which were lashed together with telephone wire from the Intercontinental Hotel in Kinshasa, and, in one flawless movement, he was out the back door and into the wide corridor where I was sitting, trying desperately to appear nonchalant. Then in another seamless movement he had stuffed his booty beneath my bags and, without hesitating, had run down the corridor towards the gates, the two officers in hot pursuit.

I stared straight ahead at the back of the bar and carried on smoking. 'See that?' said the barman, nonchalantly wiping a glass.

'Nope,' I replied. 'What?'

'Some guy just ran past and stuffed something under your bags.'

'Didn't take anything did he? The bags look okay to me. Could I have another Scotch, please? No ice.' Hunter and the officers had disappeared from sight. God knows what would happen now.

The officers were returning and I stared at them, sipping

my whiskey. They were sweating and looked puzzled. I stayed put and rolled another cigarette. Then I saw Hunter strolling back with a cheap canvas golf bag over his shoulder. He had removed his white floppy hat and stuffed it in his pocket.

'Wanna beer?' I asked, then murmured: 'What happened?'

'Later, Ralph. Later.'

'You guys want another drink? I'm just off to get some more ice.'

'One for the road, barman, thanks,' and I gestured for both of us.

When he had gone, Hunter said he had stopped in a telephone booth, pretending to make a call and then when they had gone past, he had slipped across to a travel bag shop. He had seen this golf bag and calculated that it was the perfect size for the tusks. By this time the officers had given up and made their way back to the storage room and tables. I doubt they knew exactly what had been taken.

Having stuffed the tusks into the bag, Hunter slung it over his shoulder as carry-on luggage. He then helped me to take my bags down to the transit lounge to wait for my flight back to Heathrow, which was due in about an hour. His own bags were already checked through to Aspen.

Hunter got those tusks back to Owl Farm, where they took pride of place over the open fireplace. Later, he received a letter from the Customs administration, concerning a pair of tusks. It was a demand for twenty-six dollars in duty for bringing them into the country. No talk of prosecution, no suggestion of theft or wrongful acquisition – just a paltry claim by the Customs services that duty must be paid. Sometimes Hunter never listened and always feared the worst. They had merely taken possession of them until such monies were paid. 'Impounded' simply sounded like a better word and, of course, it made for a better story.

I received another letter from him. He was coming to London and could I find him a decent hotel? He needed to find 'the nub' of the story. I got him a room at Brown's Hotel, a very smart place just off Piccadilly, and I was to pick him up at the airport. Living in Fulham made it an easy thing to do. Half an hour to Heathrow. No problem except for the 6.30 a.m. flight arrival, and when I met him, a growling journey into town with him pleading,

'What I need right now, Ralph, is a hot bath, some drugs – and a warm body to hold on to.' Not a problem for the first request. The other two were going to take a little longer. 'I've done my drawings, Hunter. How's the writing coming along?'

'Fuck you, Ralph, I haven't even started yet. Please ask around a bit and, remember, we need to set up an interview with David Frost. He knows things I would like to know. But first things first!' The interview with David Frost was set up at Frost's house. When we arrived, Hunter held his tape recorder in one hand and the plug in the other, more interested in finding a socket than in introducing himself. His first words to Frost were: 'Hello, David, hang on a minute! I'm looking for a socket for this machine. So don't say anything yet, David. I need to get it all on tape.' He proceeded to climb over the furniture, looking in all the corners to the bewilderment of Frost.

Later I left him at Brown's and started contemplating where I might locate the extra items that Hunter had asked me to find, cocaine and marijuana, as if I were going shopping for a bag of groceries. Strange how it seemed such a commonplace request. We were living in a Gonzo time warp. Somehow things would get organized with a few enquiries and

a brain rattle. The Chelsea Arts Club popped into my mind and that became the place to set up a rendezvous once I had located a dealer.

In the event, Hunter met up with the dealer in the chess room behind the ladies' bar at the club. I waited outside listening to the muttering and snorting going on until Hunter stormed out of the room and said: 'Don't ever introduce me to a geek like that again!' It was later that night that, after a few phone calls, I found what he was looking for in the most unexpected place, through a girl I knew from Bernard Stone's bookshop. I drove him, slumped in the back of my car, to her house and she was waiting outside. As we drew up Hunter saw her and with a whoop banged open the back door of the car with his feet. He leapt out. Apparently he stayed all night, having found what he wanted.

But our story never got published in *Rolling Stone* – or anywhere else for that matter.

FEAR AND LOATHING ON THE ROAD TO HOLLYWOOD

1977

*Hunter meets two movie directors . . . Film crew learn
about guns and Gonzo . . . Our star is sick . . . On the
road and heading for Barstow*

I was not to see Hunter again for another three years, when
I returned with TV director Nigel Finch to make a film about
our collaboration for *Omnibus*, a BBC documentary arts
programme. Nigel rang me out of the blue and asked me how
he could get through to Hunter and meet him.

'Send him a telegram,' I said, 'but make it very direct and
urgent, or he won't respond.'

'How do you mean?' he replied.

'Send something like "Arriving tomorrow STOP with a full
film crew STOP Ralph arriving too STOP Will need guns
STOP OK."'

Within half an hour Hunter was on the phone to me.
'What the hell's all this?'

'Great!' I said. 'The BBC want to make a film of us going
to Hollywood and stopping off in Las Vegas! Seems like a good
idea.' I filled him in on the details and Nigel phoned him to
explain his idea. It was definitely a goer and within twenty-four
hours, Nigel and I were on our way to Colorado, without the
film crew, who were going to follow on a couple of days later.
I was looking forward to seeing the ole bastard again.

Nigel Finch was a gentle soul, easily given to laughter,
trusting and extremely conscientious about making films. This

was his first film and *Omnibus* was the most serious arts programme of the late seventies. To be in it, let alone make it, was a feather in anyone's cap. We enjoyed our trip over and Nigel made a few notes about Hunter's relationship with me – how I put up with the insults and how it was all part of the creative relationship.

It certainly was but I hadn't actually seen hide nor hair of him for three years. I was as excited as Nigel. I wondered whether our chemistry would still be intact or whether we would we be like strangers again?

We touched down in Denver, which at the time was still a hick airport, small and made of four by two wood framework and hardboard. The plane that flew us over the Rockies was a twin-engined Dakota that looked like a remnant from World War II. It was part of the fleet called Rocky Mountain Airways. I loved it. The ride was bumpy and sometimes downright erratic, but in a certain way you knew that the old crate had flown over the peaks so many times it would practically fly itself. I still have the photographs of the actual flight. Someone on board took pictures of us and it passed the time on the last leg of our journey as Nigel began to get nervous.

'Don't worry, Nigel,' I reassured him. 'When I first met him he scared me, but inside he's a pussy cat – a dangerous one – but nevertheless a clumsy, gentle person. He is very gracious to his guests. You'll be fine.'

'I think we are coming in now,' said Nigel as we started the steep descent and the whip alignment into the valley to approach Aspen airport. We stooged in as always with a pilot at the controls who had done the same thing a million times. We taxied in and arrived at what reminded me of aircraft landing in old black and white movies, Howard Hughes and all that, the stuff of aviation history. Aspen was the archetypal airport I always think of as a landing convenience, something familiar and welcoming, chickens on the runway and relatives waiting for loved ones. No endless taxiing to find a vacant gate and no customs harassment or long queues of people suspected of villainy and bombs.

Our bags came off the aircraft in minutes and we picked

ours up, moving through the homely airport lounge to the front of the building. Nigel went for a pee and I stood out front against a brick buttress, one of the few that supported what was mainly a wood construction. The sun was shining and it was quiet, an atmosphere of suspended animation. It reminded me of the scene from *Bad Day at Black Rock* when the one-armed Spencer Tracy descends from the train onto the platform and waits as the train pulls away. All is peace and tranquillity. Nothing stirs but a gentle breeze.

Nigel was just emerging when there was the roar of a high-powered V-Twin engine. Nigel froze as he watched a bright red Chevrolet roll into view, rev sporadically, as though preparing to attempt a chicken run, pause for a moment and then lurch forward straight towards me. At the wheel was the familiar figure of my friend, HST, driving straight for my legs as I stood with my bags against the brick wall. I guess I should have leapt out of the way, but such was the trust, the faith, I put in my old friend, that I merely stood there trying to roll a Golden Virginia cigarette, and waited. Hunter jammed on the brakes with moments to spare. The car stopped an inch in front of my shins and Hunter leapt out over the door, saying: 'You're late!'

'You bastard, Hunter! That could have been nasty.'

'I know,' he said, 'but you knew what the odds were when you decided to fuck with me.'

Nigel was a little flustered by this peculiar demonstration of bonhomie. I introduced him to Hunter, who welcomed us both warmly after his little joke.

'God, Nigel!' I said. 'What a pity the film crew hasn't arrived yet. What an opening to your film.' We agreed it would indeed have been a powerful moment to start and a couple of weeks later when we had all but finished filming *Fear and Loathing on the Road to Hollywood*, Nigel attempted to re-enact the moment, but it was gone, the spontaneity was lost and he had exchanged the Chevy red shark for a red Volkswagen Beetle Coupé.

Walking Nigel down the front drive of Owl Farm, having invited us to the Woody Creek Tavern right away, Hunter was talking seriously about guns: 'Like this one here,' he said,

suddenly whipping a Magnum .357 out of a holster and firing it three times into the air. 'That's what I mean,' he was telling Nigel. 'One must be prepared at all times.'

'But won't someone hear that and report you to the law?' said a concerned Nigel.

'Of course,' replied Hunter, 'but I have an arsenal of them in a locked-up safe. No one would dream of it. They will have heard that fucker down in Aspen and know that all is well. You're safe here, Nigel, and prepared.' Reassured, but uncertain, Nigel kept the uneasy grin set on his face and from then on knew he was in for an uncertain ride, but also a great film. He always remembered that part. Nigel's film encapsulates the early halcyon days of an Aspen for drop-outs and hippies, but also captures the essential outlaw philosophy that marked Hunter down as someone with his own style of destiny.

'Told you you would be safe, Nigel.'

In the film, Nigel quizzed me about Hunter's love of guns.

Nigel: What about the violent side of him – the firearms, the Mace, the aggression?
Ralph: Well, as far as I know he's never shot anyone! As far as I know, he's never hit anyone. I think that's

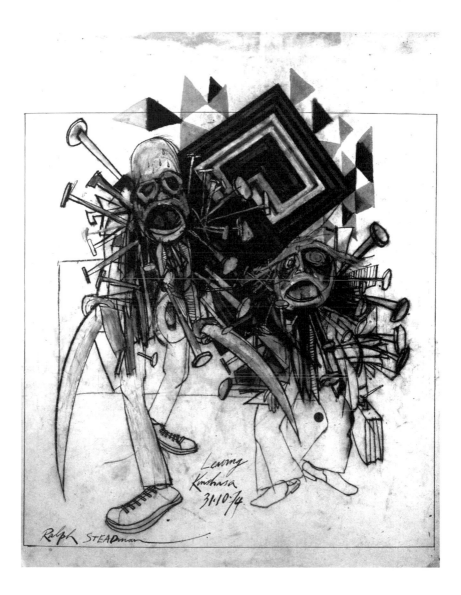

1974. Unpublished drawing of Hunter and Ralph as voodoo dolls intended for the never-published *Rolling Stone* issue on the Ali–Foreman fight, described as 'the biggest fucked-up story in the history of journalism'.

1974. Intercontinental Hotel, Kinshasa, Zaire. Hunter relaxes on hotel bed before the Ali–Foreman fight, known as 'The Rumble in the Jungle'.

Hunter in hotel pool.

Ralph in hotel room.

Hunter negotiates purchase of elephant tusks with $300 of non-negotiable American Express travellers' cheques.

1974. George Foreman press conference, Zaire.

Hunter's study of Ralph levitating in pool at Intercontinental Hotel, Zaire.

Voodoo doll, Zaire.

Hunter phoning America from Zaire.

1979. L'Hermitage Hotel, LA, during a rewrite of the introduction to the film *Where the Buffalo Roam.*

Ralph fools around in front of his graphics.

Hunter manipulates electronic machinery in hotel room.

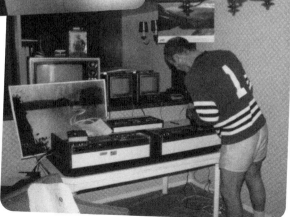

1979. Studies of Bill Murray during the filming of *Where the Buffalo Roam*.

1981. The fishing tournament, Kona.
Ralph in the fighting chair on the
Haere Maru.

Ralph and Wild Turkey.

Hunter with son Juan on board the
Haere Maru.

Captain Steve, owner of the *Haere Maru*.

1981. Hunter on board the *Haere Maru*.

Hunter talks to Captain Steve.

Hunter, deep in thought, considers our dilemma.

Hunter reads the paper during the lull.

All at sea being towed in by the *Little Child*.

1981. Drawing area set up at Owl Farm to complete the drawings for *The Curse of Lono*.

1980. Hunter watching runners at the Honolulu Marathon.

'Nuclear War – Race Cancelled' – Ralph's chalked message to marathon runners at party held at the bottom of Heartbreak Hill.

all part of the Gonzo spirit. The spirit of Gonzo is in all these implements just being used, they're all there to use and they're all parts of the same fire and punch and intensity that his words are trying to get over. He probably needs those things but they're all shot off into outer space. He might shoot them into the hillside. I don't think he's even Maced anyone, except me, of course, although he's talked about doing it. That's the other thing – Gonzo is controlled madness. It would use anything within its range to activate something. There's not really anything sinister in it. I mean, I'd be very unhappy about some of the people I have met who would tote guns and, you know, you'd feel that they were doing it with a certain malevolence, which is certainly not the case with Hunter.

Nigel: What have the two of you got from working with each other?

Ralph: Well, out of it I got some of the best drawings I've ever done. It seems like a masochistic thing, you know, to go on these assignments and end up like a washed-out rag and punish yourself. But, you see, often I can't start without being in a certain frame of mind on some of these things, there's no energy there to do it, to really get down to it, to be incisive and aggressive on the page. It's a better way of doing it than actually going out and beating up old ladies, isn't it, this type of journalism? It does get rid of all my aggression. Gentle as I appear, there are odd moments when I feel like breaking something. So, drawing's quite a good therapy, from that point of view. And I think Hunter's got that – the use of these arms – he loves all those things. I was in a gun shop with him in Washington and he was actually quite horrified by the man's obviously demonic enthusiasm for the guns and the things he was showing Hunter. Hunter just wants to see it – it's a fascinating area of life and it does exist. There's a strong gun lobby in the United States and it's partly to do with money, and partly to do with that old traditional idea that guns were part of the Old West, part of the American way of life, for protection, ostensibly, but they have somehow become something else as well. They're rather sinister and this shop was just loaded with stuff and the guy said: 'Well, you really should see what I've got in the back!' 'What have you got in the back?' asked Hunter. 'Well, I've got these amazing exploding devices which go off just in one little area like a little bang that doesn't spread, so that you can injure people by just throwing it at them. It'll explode on impact . . .'

Hunter's reputation, by this time, ensured that he had access to every major political, sporting and social figure of the time.

'All I have to do,' I said balefully, 'is see everything in pictures that he saw in words.' I felt like Edward, Hunter's mynah bird, back then. He tormented me as though I were in a cage. He was my hope and my misery. I wasn't sure what to expect at any moment. It was my innocence that saved me. And he knew that. He would give me leeway and then he would tie me down to a set programme of rules that would have purpose and logic, but not my own.

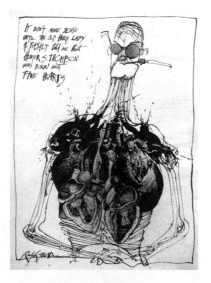

There are no ways to explain just how the film came to get made, except that I would put it down to innocence, perhaps Hunter's, but, certainly, Nigel's. To watch Hunter as he was then, in that thirty-year-old film, is to watch the flesh and bones of a very real, naïve American pioneer, full of uncertainty and yet the sense of wrong that was being forced onto the people. His hesitant way of expressing anxiety and rage, his certainty and his suspicion that what he was saying was and is right, drove a stake through the hearts of every red-blooded American who wanted to believe in some real American Dream. Even if the Dream wasn't perfect, it represented the very most that any human creature searching for a Dream could realistically expect.

What Hunter was experiencing was nothing more than national pride, a sense of himself as an American, that could be transfused into any one of his fellow countrymen if they had the courage of their own will to succeed and drive the idea forward. That spirit became the essence of a desire on his part to truly expose himself and expose the lie of American life. Hunter was the lie because the lie represented everybody. Every louse, every bestial schemer, every low-down scumbag, every sick wank-sack drove their nails into Hunter's vulnerable skull and weakened that part of him that was his strength. But he was still young enough to withstand the pressure. He proved it when the pressure was on him in the film and he threw it back at Nigel and the film crew who were at this moment tormenting him like he tormented Edward in his cage.

I don't think I was any part of Hunter's plan, except as the visual bridge that explained to those who were not going to get it that he was determined to be a one-off, a raging loner, an example to anyone who wanted to try and live with the same determination, that there was an opportunity for anyone to exorcise their old selves and start again like an unidentified prole.

It was because of his example that I decided to become an American. I filed papers, registered my intention to adopt the American flag, swear allegiance and take the Oath. I told Hunter this and he said: 'Ralph! I will do everything in my

power to prevent you from ever becoming an American citizen!'
I was shocked and mortified that he did not think me worthy
of such a station in life.

'I thought you would be pleased,' I said.

'Pleased?' he snorted, 'Pleased?' he said again. 'You can
never be an American, Ralph, because you are Welsh!'

'There are lots of Welsh Americans, Hunter,' I replied.

'Except you, Ralph! You are far too weird to even consider
the honour!'

'I lived for the possibility. I thought you would be proud.
I thought you could pull strings that would make it a warm
possibility.'

'Cold helplessness, Ralph! Cold helplessness! Never!' It was
then I knew I was a pariah in his life. I never understood why
he hated the idea so much, but he did hate it.

When I began this book I thought it was going to be a journey
of pleasure and warm memories, but as I write I feel more
of the icy winds of rejection that were probably there from
the beginning. There is a point at which nothing was ever
worth the effort, nothing given and nothing taken away. My
involvement was nothing more than my own ambition. Quite
by chance I became a part of this man's life, more as an infec-
tion than a friend. I fooled myself that there was something
in me that he found important. Actually, as time went by, he
hated the very idea that something as putrid as a cartoon
drawing could ever capture the essence of what it was he was
trying to describe. But when I search deep inside myself,
grasping at words like air, I believe he may have been right.
There was no purpose in my involvement. Why should we
ever have known each other? Why did that happen? It wasn't
even his choice. I was the flotsam that *Scanlan's* gave him.
Stuck with me, he persevered, not unlike a benevolent uncle
who was landed with a young nephew for the day. He was
kind, he was even philosophic and he was generous. I was
invited into a life that was as incomprehensible to me as a
glimpse into an exclusive club for gods only.

The film crew arrived and for two or three days were getting to know one another like a bunch of animals in a zoo or, to be precise, were running around Owl Farm with guns, apprehensive and a little scared, absorbing the infectious presence of Hunter's work and lifestyle until it became almost normal.

They would never know the whole Hunter. Nobody ever did, though I do believe that Juan and his mother Sandy fought his corner, nurtured him even, and protected his right to be lame. Many did not understand, for there lies all his weakness and his shame, the 'panicked sheep' syndrome, of an American who will not be penned up and left helpless. His aim was not to show his wounded knee or even his pride. He would show them something of himself, but not everything. That would be privileged information. So far, these interlopers had only fired his guns and already given themselves some footage — useful footage too — that would end up as part of the film. So who was doing the work so far? I had already been interviewed and spoken some of my truth, but Nigel's film had to traverse a minefield and on this first lap it was all a bit like summer camp. Here was his crew dragging heavy film machinery around, eager to get going and take part in something, apparently steeped in fun.

Nigel's instinct was to make for Las Vegas, surely the centre-spread for anything to do with the heart of the man he was trying to capture in the most seminal framework of his film. It was Nigel's big break so there was no question of failure. Anything that happened had to somehow embody the soul of the subject of his film. It was time to move out of Aspen, get on the road and bite the bullet.

Nigel and I stayed in a holiday cabin for the first few days and swam every morning in its private pool and, during this swim, he asked many nervous questions. At the same time I had a drawing to do for a magazine whose name now escapes me, although it could have been for *Penthouse*. I remember the drawing well. It was a picture of double open-heart surgery, with a head at either end. 'That's what I want my film to be,' he said.

The picture was finished before we left the cabin and maybe Nigel kept the image in his mind as something to aim for. I wasn't to know that at the time, but, with my help, that was exactly what he was aiming for. He was clutching at anything that would give him a handle to hold on to. I didn't fully appreciate then that I could help by being his navigator. I was there as a bridge and as some friend, both Nigel's and Hunter's. I was to be the 'innocent' catalyst. Looking up the word 'catalyst' in the dictionary it said: 'noun: a substance that increases the rate of a chemical reaction without itself under-going any permanent chemical change – a person or thing that precipitates an event'. That seems to have been my role all along. The more involved I became in the act of a Gonzotic event, the less my presence was required, because Hunter had found what he considered to be his very own true territory, his indelible identity.

We were now on the road, making for Las Vegas, filming on the way and, most important of all, stopping off at what is now one of the oldest hotels in town, the centre of *Fear and Loathing*, Circus Circus. Today it is a run-down version of what Las Vegas has become, but in 1971 it was state-of-the-art American Dream. The Golden Nugget, the Fremont, the Tropicana Hotel, where we stayed, the Stardust and the Lido. But to stay there for long was suicide. We needed to get out on that road to Hollywood across Death Valley, through Barstow.

The roads moved out in all directions and turned into sand. They were roads but they were more like dirt tracks, which is natural enough, since the entire fabrication of Las Vegas rose out of a desert landscape. The signs enticed you to go inside and spend and I would have been stupid or insane if I thought I could function in such a situation. But this was a stopover. My abiding image at that time was waiting at about midnight to take the elevator up to my room, idly watching a row of old biddies sitting in front of one-arm bandits, their plastic beakers of change and drinks by their sides. They were there for one thing only – to feed off the tit of Mammon – maybe to win something, even briefly, and then go home to Des Moines, Cleveland or wherever, wasted and broke. I groaned at the weird emptiness of the scene around me before entering the lift. I was glad of the sleep since I rarely got any when I was around Hunter.

I rose around seven and went down in the elevator to get some breakfast. Incredibly, the same biddies were still on the machines. A couple of them had fallen asleep against the hard chrome metal, too tired to move, dreaming their American Dreams of pots of gold at the end of a long curved neon sign with an arrowhead which said: 'WELCOME'. There must have been a reason why I put up with the pain – looking for my pot of gold at the end of my own rainbow, maybe. It was a dirty sidewalk, I didn't see any names and I didn't want to see any names. I wanted to hide. Hunter railed and played the game of the plagued victim of something he thought was a good thing to go along with at the time. I knew what I meant and I knew what I did 'and so did Nixon' was his reason.

It was the day before we took off for Hollywood, before we crossed Death Valley and began our descent into Los Angeles and made the grand entrance along Sunset Boulevard and before the doubts flooded in to cloud Hunter's judgement about Nigel and his competence. I talked flippantly about playing a joke on Nigel. It had to be something about his 'star' that would shake Nigel to the roots. I managed to get the crew on one side, without their director, and talked to them about creating a scenario that would put him into a cold sweat.

Nadia, his continuity girl, would inform him that something was wrong and he should call Ralph. Hunter loved using make-up, grease paint, lipstick, and wearing wigs. They were props for a man who always wanted to make a scene, a gross mismanagement of what would otherwise be theatre, the elements he loved to hide behind. Things had gone smoothly and a sudden jolt might just generate more focus on the content of the film and the direction it was going.

The plan was very simple. Nigel's 'star' would appear to be in a terrible, feverish condition. He would be sweating like a pig, vomiting blood, and on the phone making 'final' arrangements for the dispersal of his estate, talking in grave tones to his editor, Jim Silberman, in New York, about the final edit of his book of collected works, *Songs of the Doomed*. Hunter had wanted to make something happen that would scrape gristle off the bone, make Nigel 'pay' for his story. And so the sting was set up. The crew would hide behind the curtains in Hunter's room and Hunter would make up to look horribly sick and feverish, as though he had a serious bout of food poisoning.

I waited in my room for a call from Nigel which I knew would come because he couldn't find his crew. The phone rang. It was Nigel.

'Ah, Nigel! What's going on?'

'That's what I want to know. I can't find my film crew and I am getting groaning noises from Hunter's room. Can you come down?'

'Okay!'

I knock and hear a groan. 'Hunter . . . er? Are you getting up now?

'I'm getting up Ralph, oo-oo-o-h.'

'You are?

'Yes.'

'Okay. Um, how long do you reckon you'll be? You all right?'

'O-oo-oh-All right.'

'You don't sound too good to me.'

'You got people with you?'

'What?'

'You got people with you-o-oo-oh?'

'Well, just me and Nigel. We're just thinking about checking out, you see.'

'Oh?'

'Yeah.'

'Where we going?'

'Well, we gotta start on our way to Los Angeles, you know, so er . . .'

'You wanna go now?'

'Why, do you feel sick or something?'

'Yeah. Fucking weird.'

'Well, er, shall we come back?'

'Please. Give me some time to compose myself.'

'Okay, righto, Hunter. I'll come back in a short while.' Hunter groans convincingly from the other side of the door.

'What do you think is the matter?' said a concerned Nigel. 'And where's my film crew?'

'Oh, I guess they'll be gambling and just waiting for you to give the signal. Hunter will be okay. Give him an hour. I wouldn't mind betting he is just making you nervous to keep you on your toes. Let's get a beer and then knock again.'

We returned and I knocked again. The conversation that ensued went something like this:

(Knocking.) 'Hunter?' (Knocking.) 'Hunter? You awake?' (Knocking.) 'You there?' (Knocking.) 'How do you feel today?' (Knocking.) 'Hunter? Hunter? Hunter?' (Whistles.) There is a groan from inside.

'What's the matter — what *is* the matter with you? Come on!' There are more groans.

'Open the door!' Further groans.

'Oh, my God! What the hell's wrong with him? And you haven't found your crew, Nigel? What about your continuity girl, Nadia? Can't you get her on the phone?'

Hunter calls out some strange cat wails . . .

'I'm hearing all sorts of awful stories, Hunter? What's going on?'

Hunter groans again, overdoing it as usual, I thought to myself. I wring my hands and play distraught. 'Oh God! Did anybody see him? Nobody saw him?'

As I am saying this, the door is unlocked and I fall into the room.

'What's the matter with you?' I catch sight of his face and it is greasy, ashen-grey white, and there are huge beads of 'sweat' all over his head. His make-up looks smeared but horribly gross and strangely convincing. Christ! I thought, I know what he's done. He has doused himself in water and he's soaking — even his shirt.

'I'm okay. I'm all right,' he keeps mumbling and then says: 'Well, actually, I think I'm dying.'

'That's good. You haven't got leprosy have you? God! You sure you're gonna be okay? You look shocking to me.'

'I feel it,' he growls and holds his stomach.

'I think you look terrible. Definitely more terrible than usual. Hey, sit down, sit on something – on anything, but just what are you doing . . . ?'

'I'm on the phone, trying to phone my lawyer.'

'Oh, you're well enough to do that? Well that's good.' Nigel has followed me into the room and his face is a picture of abject misery and concern. 'He's trying to phone his lawyer,' I said, but I don't know why – yet.'

'Yes, those numbers again, they were . . .' Hunter is holding himself and speaking with an agony growl.

I speak to him again. 'Can you, um – Hunter, speak to me . . . Nigel's lost his film crew and . . . good God, man, you're sweating like a pig!!'

'Been vomiting blood too, this morning. Yes. Yes,' he says talking down the phone again. 'Gimme, gimme the numbers. Give me Silberman's number. No, well both. Two, four, six. Two, four, seven.' He is ignoring us a moment while he speaks down the phone.

'What?

Hunter is playing up now. 'Oh, aarrgh!'

'Do you want anything? Say a few last words – anything, you know . . . if we can find the film crew and we'll get it on film.'

'What's that?' Hunter says again down the phone.

'He looks pretty desperate to me,' I say, turning to Nigel.

Hunter goes on talking down the phone – 'Three, four. Three three five eight. That's a doctor? The doctor. Yes. Tell him no. Yeah. Tell him I'm dying. Four three two one. Well, he'd, it'll be that one but. Just a minute three two, I'll put it down. Mm. Okay. Hunter pauses again and holds his stomach Then he gets up and retches over the wash basin. 'Four five six. Nine o seven nine.' He pretends to write more numbers down. 'What else do we need? Four five six. Six seven seven. No, hell I was getting all ready to take off . . . then this! Bookman? Book . . . yeah. No, should be up at that country place. Two o three, two, two, seven, eight, nine, one, six. Bookman? A home number? That's . . . my book though. That won't be until tomorrow. Gimme Bob, Gimme Bob's home number, yeah. Quick!'

'Can you last that long?' I say. 'Hey? Can you last that long? You look terrible. There must be a hotel doctor.'

'Yeah!' Hunter continues to talk down the phone, talking numbers. He fumbles and searches. 'I had it in another book though. Sorry! Ribbleton. Yeah. Six, five, five, eight. Okay. Er. Bert Schneider. Hmm. Both. That's in the book anyway, Later. I'm very sick now – I've been vomiting blood. Okay. Two, seven, four, five, five, oh, seven. That'll do. Right. No. Yeah. Hm? Oh, helped a lot. Oh yeah, yeah. Yeah. Er, you can speak to Ralph for a second here, yeah.' As he hands me the phone he says wincing: 'Tell Sandy just exactly what's been happening.'

I take the phone and say: 'Hello. Sandy, it's me Ralph. Hi. Well,' I continued, 'he looks sick to me. Nigel? Oh, I think he must have gone to get a doctor. I'll keep in touch.'

'D'you have the snack by the way?' says Hunter. I have forgotten all about the snack. 'This . . . is . . . awful . . . o-oo— oh . . . this feeling of drained blood. I could use a crab sandwich.'

I have to suppress a smile and look away.

'A terrible thing, um, here. Yeah, yes. O-ooo Oh!'

'Hunter, surely you aren't dying – not if you feel like a crab sandwich?'

Hunter starts to laugh. 'A terrible thing!' Nigel has come back into the room, as much worried about his film crew as about the state Hunter appears to be in.

'Ah, Nigel! I dunno, it's er . . . I've never seen quite . . . I think it's er – pleurisy. Is that lethal?' Nigel looks as sick himself now and still has no idea where his film crew has got to. Hunter starts groaning again.

'I don't recognize anyone any more. My vision's gone and . . . don't lie to me . . . oo-o-o-oh!' moans Hunter.

I try to reassure Nigel again, but he is the sickest man in the room at that moment. I keep up the charade but I can also hear repressed titters from behind the curtain. I speak a little louder and say: 'I'm hoping, I'm hoping he'll say a few wise words to the world before he goes because it's er, it's er it's horrible – the thing is you shouldn't be drinking beer, Hunter,

you shouldn't be drinking at all! You should be having peni-
cillin and er, you know, blood plasma and things like that at
this moment. Otherwise if you don't I think it's all over — look!
he's sweating like a pig, again.'

'Phew! Fuck the poor,' Hunter groans into his pillow.

'Oh, that sounds good? Oh. Thank God. I — you're . . .'

'Fuck the poor,' says Hunter into his pillow again. He seems
to be shaking with fever.

'God!' I say, turning again to Nigel. 'I mean, he looked a
bit weird at four this morning — he looked a bit weirder at
four-thirty, but I just thought that was just er, you know, er,
well, playing around, but it's obviously not, it's . . .'

Nigel's voice is trembling. 'I know this is tragic but I am
so puzzled. I mean where is my film crew and not only that,
but . . .'

Hunter groaned again. 'I don't recognize anyone any more.'

Nigel continues: 'What about my film?'

Suddenly Hunter is sitting upright and fierce. 'WHAT
ABOUT YOUR FILM? Your film!! Fuck your film and fuck
you and your rotten film crew! I am at death's door and all
you can think about is your goddamn film! You weasel!! You
cock-sucking bastard. I am sicker than a mother without tits
and all you can think about is your fucking film!'

At that moment the film crew leapt out from behind the
big damask curtains. In spite of the tirade from Hunter and
in spite of the possible demise of his 'star' I never saw a greater
sense of relief on anyone's face or a greater sense of embarrassment as there was on Nigel's face as he realized that, in
his consuming desire to capture the very essence of Gonzo, he
had been taken in and flogged like a stray dog. From that
moment on, he stalked his prey as one would stalk a wild
animal — by giving the beast its own space in which to operate
and following its tracks. Just for the record, Nigel edited most
of that scene out of his film, but left in just a short sequence
of it — a token. It makes no sense at all.

The cars were hired and were brought around to the front of
the Tropicana, where all the gear was loaded.

'We're off! We've got a beautiful car. I don't think he'll make it though, I mean, we've had to get him well, so we've got one with a stretcher in the back . . . just in case, Ha! Ha!'

'Hunter can lie out. I think I'll have to drive. It's a Courier, a Dodge Courier . . . Oh, dear!' Hunter hated Dodge cars and a replacement at short notice could not be found – not even a Dodge Aspen, which would at least have been appropriate!

Hunter was recovered.

'I feel better now, Ralph,' Hunter said. 'Let's go to Winnipeg!'

'Well,' I said, inspecting the car, 'it looks nice inside. Beautiful and brand new! Anyway we've got to get on the road. Winnipeg here we come!'

The crew were ready and we started and then stopped. Stopped and started. Hunter cursed the car and then the film. Then he cursed me, for getting him involved in the wretched film idea in the first place. He blamed me and complained of my cupidity and my mendacity. We had a sound recording man called John in the back of the car, ready to capture words of wisdom, Gonzotic speech patterns and anything that might add authentic touches to the film.

We stopped again about half an hour out of town. Hunter wanted to film the desert with the neat little BBC camera they had lent him to film anything that took his fancy. 'Technology is still not up to it. Reality, I mean. Machinery like this warps reality,' he said. The crew, of course, got out of their cars and set up to film again. Lots of footage of Hunter walking, peering into the distance and the bleakness. We were at a crossroads that went east, north, south and Gonzo! Then we were off again.

On the road we tried to generate a serious discussion . . .

Ralph: See, the thing that fascinates me is the reason, the actual, the *real* reason for trying to twist anything at all, to try to make it work. I mean, it's about twisting things. It's about cranking things up in a way that it's not like going somewhere and trying to factually report and say: 'That's what happened.' It's just making something happen in order to write, otherwise there's

nothing to write about. It's just what comes first, the
chicken or the egg, you know. Where does it begin?
Hunter: Money.
Ralph: Ah, money — it's all money to you. It offends
me to hear you say that, you know. I really do feel
offended by that, the whole idea of . . .
Hunter: You wretched, half-mad Judas-goat!

In the film, Nigel tried to get Sandy, Hunter's first wife,
to explain the difference between English and American
humour and in particular, Gonzo humour.

'It's closer, do you agree, do you know what I'm talking
about,' she said. 'It's more bizarre and it's not quite as "Ho Ho
Ho", you know. "That was funny" and you get a slap on the
back. American jokes are more laid-back than that. But Gonzo
is deeper, more visceral.'

Nigel suggested that Hunter's sense of humour was at times
quite strange and bizarre. Sandy agreed and he added: 'And
it's pretty cruel too.'

'Yes it is,' Sandy said, 'but not always — I mean not always.
It is and so are Ralph's drawings, very often; it's a very cruel
humour. But you notice it right?' Nigel asked if she had any
observations about the way Hunter and I worked together.

'The point is,' she said, 'I really think that they love each
other and I think they work beautifully because they balance
each other and I think they work very well together because
I mean they are both a little crazy, but they are also very strong
people and I know that Hunter has helped Ralph out and I
know that Ralph has helped Hunter out very often. It seems
to balance, and of course their humour works together. They
have a good time, you know, like kids, I'm not sure which one's
older, I think Ralph, yes Ralph! About twenty years older.'
(God! That was a bloody cheek!)

Nigel then asked her what she understood Gonzo journalism to mean.

She could not be sure but she felt prejudiced in different
ways. For Sandy it meant a looseness that she wouldn't want
to see in others' work. She liked to see Hunter's work tighter

and she thought that his tenth anniversary piece about Oscar Acosta, *The Brown Buffalo*, was what she wanted to see more of. She preferred his work to be 'loose' and 'wild' and 'not tied together'. She liked the spontaneous wisdom, madness and wildness. To her, Hunter was much bigger than just Gonzo. He was a Southern gentleman and very conservative in a lot of ways, with which I would concur. She also saw him as a wise and compassionate man who could, quite suddenly, be a child.

I would say exactly the same. Hunter displayed soft moments that suggested some early, hurtful abuse in his childhood that made his behaviour patterns erratic and vulnerable.

Nigel wanted to get some idea of how Juan, Hunter's son, felt about his father's work.

One can tell from Juan's reply that, as with my own sons, there was a reticence. There always is. He had 'started' to read *Fear and Loathing in Las Vegas*. He had got about halfway through, he admitted on camera. Then he started to read the next book, *Fear and Loathing on the Campaign Trail*, but he admitted that, frankly, he was more into science fiction than 'that type of writing'.

It was very hard for Juan to say exactly what he thought about his father's writing. He didn't want to say the wrong thing on film, obviously, but if someone you are close to does what they do all day long, or all bloody night long, as in Hunter's case, it is a bit like asking what you think about your father's breathing.

'Well, er, it's hard to say . . . I mean, I don't . . . er . . .'

What could Juan say? His father did it every day but Juan had other interests. If I were Hunter's son, I believe I would also put some distance between me and him. The goddamn house was a writing factory.

Nigel wanted to know whether Hunter was like his books or were the books like Hunter. I think in some ways there was a moment when Juan began to wander off into his own protective world. His answer was ambivalent. He said that he thought that Hunter was the way he was and he was unloading his mind like a dumper truck. He was up all night and on drugs constantly. It must have been tough having a raging beast for a father, gentle or not.

Crossing Death Valley, Hunter speeded up to test our bucket of an automobile, moving through the gears, automatic, but still being operated by him, and swerving, as if he was driving a bumper car, showing off for the film, of course. Always for the film. In spite of his protestations and his loathing of the 'filthy media' that rode with him, he rode this fine beam, this breath of light in a stone dark world, like a dragonfly searching for sunlight, that last ray, the green flash on the Pacific horizon, before it dies.

I had been with him at night driving fast into Los Angeles when he had said: 'Look at that moon, Ralph. Have you ever driven by moonlight with the top down, but with the lights out? The wind is fierce and strong. But by that light, Ralph, you can see the bats. You can see the stars but the police can't see you, and they never see the bats. If they are lying in wait for dumb fucks followed by a fairground of lights they get caught. Don't advertise. They only hear us and by then we have slipped past in the dark . . . and not only that, they would never believe that anyone would be so goddamn stupid as to drive at night without lights. It confounds reason. That's Gonzo, Ralph . . . That is pure Gonzo!'

I knew that the rental car was getting on Hunter's nerves. To him it was a piece of serious American junk. Its wheels were screeching, but not because of any fast corner work by him. They seemed out of alignment. He blamed Nigel for getting him on the cheap.

'That's the BBC, Hunter. It's an honour to work for them! Yessir! It wouldn't do to sell your soul for money, not to them,' I said. 'Get him a cheap car. He'll never know the difference. Americans don't, y'know.'

'Ho! Ho! Ralph. Very funny. But they will pay for this. It will be the most expensive ride they will have ever taken!'

Hunter hadn't bought an American car since he had been in his teens. When Anna and I visited in 1973 for the first time, the car in front of the porch and the peacocks was a Volvo. He said he would never have a German car and then turned up at

Aspen airport in 1977 to pick me up in a red Volkswagen Beetle Coupé (for the filmed sequence). When I asked him about that he said: 'Yes! Isn't that perverse? The bastards have always got a surprise up their sleeves. Who'd have thought Hitler was a socialist? Nobody is all bad, Ralph! I admired him enormously – to dream up a hideous scheme, gather his troops and see it through. He was a sick fucker who knew that there were other sick fuckers just like him who would come when they were called. How he knew it, I don't know, but the bastard knew it and practically achieved it. And while we're on the subject, Nixon saw it too, but he got stupid, and recorded it all – like our friend here in the back of this treacherous car, recording everything we say. So . . .' and he turned to address the sound-recordist in the back seat, 'would you mind deleting that last bit of chatter. Things like that get misconstrued.'

'Oh,' I said. 'Well done. Spoken like a true Son of the Kentucky Pioneers!'

'Ho! Ho! Ralph! We were ahead of you, when this country began to crawl out of the dark ages. And as for the Welsh, well, Ralph, these crazy pioneers were thinking of Democracy before you even learned to take a bath.'

We drove in silence but considered the car intermittently. We agreed that it was probably one of the first examples of a car with built-in obsolescence. Like the New Dumb, its time was finite. Nothing was going to last for ever. What we had not anticipated was that the New Dumb were here to stay and were using up the planet like a rechargeable battery. None of us knew any of that in the seventies. As we lived through that decade, we thought that we had reached Valhalla, the Hall of the Slain, but had survived. We thought we had suffered all our wars and now we would enjoy a time of peace and tranquillity. However, people never learn and a newer type of war was to envelop us. We seem doomed to repeat all our mistakes and find that we are being groomed, and doomed, by the legions of war, Odin's messengers, who would have us believe that aggression is the only way forward. Now we live in a war zone, the entire world. It manifests itself in every bug-infested hole, every fetid sense of right, every feeble-minded, belief-drenched moron, who seems to think that a bomb is a good way to prove a point.

We were somewhere around Barstow on the edge of the desert (hmmm!) when we started to go over the hump and down-hill into Los Angeles. Somewhere around Baker, I was rolling the last joint for Hunter using the heavy rolling machine I carried with me to roll regular cigarettes. We were going to be late into town but at least we had rooms at the Sunset Marquis, my favourite hotel along with the Wilshire on Wilshire Boulevard. Nadia had assured us of that.

Hunter was still mumbling and cursing from time to time, mainly about the terrible car. Then I spilled the hash all over the floor. I was trying to mix it with regular tobacco in order to spin it out. As Hunter raged against my clumsiness, I picked up every last morsel using an old 'Native American' trick – slightly damped tissue using beer.

'That's an ugly joint, Ralph.'

'It'll be perfect, Hunter. It picks up everything and . . . watch where you're going-arrgh! Sod it! I'll have to start again.' There were rocks everywhere at the roadside and this was no time to hump over them any more than was necessary. Then there were the cactus plants that drew Hunter like a magnet just for the thrill of feeling the bumps – especially in this car. He was determined to arrive with it beaten and bruised. He never forgave a cheap thought, particularly a BBC one, and he was determined that they should pay for their meanness.

I finally made the joint, which was by now full of floor debris, sand and what hash I had managed to suck up off the floor, by the simple principle of osmosis. I held it out of the window to make sure it was dry and then I handed it to

him. Hunter lit up and inhaled deeply, held it in, emptied his lungs slowly and then said: 'Hmmm! Ralph, that's . . .' – cough!! – '. . . a pretty good joint. I'll make this one last. I take it all back, Ralph. Good work!'

'It's in the sand,' I said proudly.

'What sand?' Hunter looked at me but missed the smirk.

'Well,' I said, 'the sand in the desert. A couple of grains mixed in with the joint augments the heat properties in the mix, thus giving off a more sustained burn.'

'Hmmm! That's wise gibberish, Ralph. Would you like a toke on this?'

'Er, no thanks. You have it all. I made it for you.' Then an idea sprung up to have Hunter get sick again, but this time freak out with blood, some of my red ink, but get it all on film. It would ruin the car but, unfortunately, the film crew was nowhere to be seen. Then Hunter wanted to go fast, open the door on his side and while getting me to hold the steering wheel, take the top off a bottle of Heineken in the door's framework. I urged him to slow down to perform the trick but he had done it before. A car behind us slowed down and put distance between us as they watched this mad guy opening his door to pull off the trick.

'You know I don't like speed.'

'What's the difference, Ralph. If I hit something it doesn't hurt any more at ninety than it does at seventy. Then maybe it will blow up – particularly if we can reach a hundred and thirty. It could burst into flames and we could go into a raging fireball and into one of those dry desert pit lakes and it's over. Then the fucking film crew have lost their best shot, because they can't keep up! But that's not Gonzo, Ralph! That's suicide!'

We had hit the suburbs of this sprawl of human dreams and delusions. Hunter pulled over to the side of the road and picked up the camera the crew had lent him. The sun was going down behind the horizon; melting yellow and orange clouds, burning orange-brown atmospheres into the gloom, that augment the sheer beauty of smog-ridden blue pollution, changing by the second and making us believe that all is well. Our film crew was still nowhere to be seen.

Then we were off again, down Beverly Hills, Mulholland Drive and onto Sunset Boulevard.

'Let's get straight to the hotel. I'm tired and need the break.'

'You're getting old, Ralph. Tiredness is in your Welsh blood!'

'That's okay by me,' I said and closed my eyes. Hunter was alive now.

'It's sure good to be back in Hollywood, where we belong.'

The Sunset Marquis is a Rock'n'Roll Hotel. Keith Richards stays there in one of the private houses that surround the main complex. I don't think he was there at that time but I stayed in one of those houses when I came over in the late eighties to help in the making of a CD-ROM on the life of Leonardo da Vinci for a company called Starwave run by Paul Allen, ex-partner of Bill Gates. I should have met Keith Richards then, but my cursed reticence held me back. To cut that story back to the quick, CD-ROM never really took off and was soon followed by DVD and then I don't know what happened. Just another fucked-up 'project'. I remember going to Seattle to meet Paul Allen, who took me into his private gallery to see his Mirós, Degas, Kandinskys and, right next to it, his basket-ball team's gymnasium. We sat on a long bench and gazed at his art collection. It was surreal. Surreal.

We all met in the lobby bar and talked about the way the film was taking shape. We had lost the crew back in the desert, because Nigel had wanted to film some dramatic desert shots, which I don't recall ever seeing in the finished film. It was agreed that we should wait until Hunter rose, or, preferably, insist that he stay up so that we could catch him in one of his dawn reveries.

This happened but it also made Hunter short-tempered and weird. As he was filmed, he was obsessed that the film crew was 'drawing a crowd', and he kept hiding behind parked cars or in telephone booths and alleyways. Nigel tried to get him to speak about the Hollywood area and the paving stones with the stars' foot- and handprints imprinted on them. That

concept sickened Hunter and he mumbled names like Debbie
Reynolds and Frank Sinatra. They were still dirty sidewalks
to him.

'Another dream factory where you can die on film and live
to tell the tale,' he said.

Hunter's film, *Where the Buffalo Roam*, would be premiered
on this strip and Nigel pointed out the Chinese Theater where
the premiere would be, urging Hunter to say something about
how it felt. But he wasn't interested. This was Hollywood and
Hunter was here ostensibly for Nigel but mainly to discuss a
film with another scriptwriter. We were also there to meet Art
Linson, the director, who wanted to meet me, too, to discuss a
poster for the film. I insisted that I was only here for the ride,
but I can suddenly flower and sting when the moment is right.
To Hunter, I was only here to make money.

Hunter had arranged to meet Nixon's lawyer, John Dean, at
the Sunset Marquis to tape a filmed conversation. John
brought his own tape recorder. They had a good talk. It was
a conversation about rehabilitation, reaffirmation; it was a
coming together of opposites who had found a common enemy
– Richard Milhous Nixon. I was pleased to see him again.

I first met him in 1976, when we worked together for *Rolling Stone* at the Republican Convention in Kansas City. Kansas City is a cattle town that became my theme for the series. Herding delegates about like cattle was a fun thing to do in a drawing. John had been Nixon's personal lawyer and would presumably know whatever there was to know about his boss. I got quite a kick out of knowing that. In 1976, in Kansas City, I awaited his arrival in my room and duly received a call suggesting that we meet. I thought it was Humphrey Bogart on the phone. He has the self-same lisp and deep timbre in his voice. Less than three years after the Watergate hearings which the world watched daily on television, he still had a high profile and his face was instantly recognizable. I have to admit I was excited to meet this man who had spilt the nasty details regarding Nixon's administration.

He said he was having a spot of bother with his portable typewriter, so I said I would come over and take a look. I suspect it was a neat ploy to get talking about something specific and it was flattering to think that I might manage to fix his typewriter. That I didn't is neither here nor there. I don't think either of us could have fixed it. But it was a neat way to get to know each other. He invited me back to his home in Beverly Hills, where I finished the drawings to accompany his article and I have stayed friends with him and his wife Mo – in spite of that!

Dean spoke truthfully to Hunter about his doubts, fears and his isolation during those days of testifying. He had, after all, taken on the entire administration, knew he was going down, but had stayed the course and told the Watergate Committee everything he knew. He went for broke and his future would no longer be in government. Those bridges were burned and he could hardly become a lawyer again. However, he *could* write books, which he continues to do to this day. The Prodigal's end of the rainbow. *Blind Ambition*, his book about his time at the White House, is very revealing and a must for anyone doing a thesis on the Nixon years and politics in general. He knows his subject intimately as a reformed Republican. Republicanism is an aberration.

He had no advance knowledge of how he would be received: 'no sense of who would back me,' he said. 'I couldn't under-testify and I couldn't over-testify. Where does Executive Privilege, as it's known, begin and end? I thought there was a likelihood that Nixon's top men, Haldeman, Ehrlichman, Mitchell, MacGruder, Colson, Kalmbach, Stans, Segretti, Chapin, Dwight — too many to name — would all testify to save their skins, making that about 186 counts of perjury against me if Nixon's defence could make it appear that I was lying, whether I was or not.'

Nixon's men did indeed testify against John, but because they were lying, they couldn't get their stories to fit together. It was only when Alex Butterfield took the stand and mentioned the famous tapes that the truth oozed out like a bursting boil. John's evidence (eight hours on the stand!) vindicated him. But as John said: 'I was very lonely hanging out there by my thumbs and walking with no room to move on that thin, high wire, I suspect I would have spent more time in prison. All I knew was that my wife and my lawyer believed me. Even after the hearings, a national poll proved that nothing had changed. Nixon was forgiven.'

While I was with Hunter in Los Angeles, during the making of *Where the Buffalo Roam*, he was thinking of Richard Nixon and Gonzo satire. At one point, he teamed up with Bill Murray and another actor to stage a fatuous demonstration. They stood on the street chanting through bull horns: 'Ask Nixon to pardon us! Ask Nixon to pardon us!' It was a fiendish jest to re-elect Nixon. They drew a crowd of three!

'What I would like to do is hire one of those light aircraft to fly over Watts,' he mused one day. 'It would be pulling a banner behind it saying: "NIXON IN 1980".' I laughed. That was what some call irony but what we called Gonzo. The BBC actually did try to hire a plane to carry out the stunt, but nobody would do the job. It was too scary and, possibly, too sick.

Hunter had now got it into his head to start making his own funeral arrangements and Nigel caught on to this idea as something that must be filmed.

We arrived at the Hollywood Undertakers of Reed Brothers, Tapley and Geiger, in West Hollywood, to discuss the serious matter of Hunter's monument. We were introduced to the mortician. 'Mike Reeves, right here,' he said proffering a firm, reassuring handshake, his face, a picture of professional gravitas. He invited us to watch a promotional film about caskets. Bela Lugosi couldn't have done a better job. The commentary was very spooky.

'There are two basic types of casket,' it began. 'Protective and un-protective. Protective caskets are designed to be completely sealed from the entrance of water and air, mainly constructed from steel, copper or bronze. Metals that provide for a long interment life.' It was such a Hollywood ethos to contemplate a long interment life – as though you were coming back!

'We're here today to discuss an idea for a monument,' I said. 'We're here to get an estimate. I'll show you this. This is the basic idea,' and I unfolded a plan I had drawn of what I thought Hunter would like. Well, it wasn't quite what he had in mind but I had to draw something and created a design based on an old metal Teutonic drinking cup I had in my studio; Wagnerian wings on either side.

'It hasn't been approved yet,' I continued. 'My friend here lives in Aspen, Colorado.'

'That's terrific!' said this very serious undertaker, who must have had all manner of wacky last requests.

'It's a bit like a giant eagle's birdbath,' I explained. His face remained implacable as Hunter added: 'We also want you to think of this valley of mountains which is about twenty miles across,' and he showed Mike Reeves a photo of Woody Creek. This inflexible man, this 'agent of death's hardware', sat up and I could have sworn he looked impressed.

'Is this the property that you own now?'

Hunter replied: 'Yeah, you see the mountains right close back there. Well it goes back further over the top right over the cliffs back there. I can do just about whatever I want.'

'But Hunter would maybe prefer,' I interrupted, 'something more like this, much simpler but with the fist on top.' I had started to draw another simpler design using a collection of stainless steel pipes. As I was drawing I carried on talking. 'I did this merely to give Hunter something to bounce off and say no to. Something like Albert Speer – d'you know Albert Speer?' The undertaker nodded. 'Er . . . yes. I . . . I've heard of him. But this is something else,' he said, acknowledging the new drawing emerging on paper.

Hunter continued to explain his plan.

'There's going to be a pile of rocks about a hundred feet tall, a giant chrome cylinder, conical, tapering up to the top, about a hundred and fifty feet tall. On the top will be a fist with double thumbs – the symbol.' I started to draw the fist but as I drew the thumb Hunter said: 'I've told you a thousand times for ten years, Ralph! Two thumbs Ralph – TWO thumbs!!'

'You put it on, Hunter,' I said offering him the pen.

'Yeah,' he said irritably, 'because it's going to be there in the end!' And he drew a twisted version of another thumb on the drawing to accompany the first.

'Well, it's better than no thumb, which is what you had. I can't count on my friend,' Hunter said, addressing the mortician, who had, by now, entered into the spirit of this massive egotistical plan.

'Well,' he said, 'you mean that this is something you plan to make after you die, is that what you're saying?'

'Oh yeah!' said Hunter. 'It's in my will. It's going to be a little hard to describe, particularly with me gone. It will be a nice monument. After I have been cremated, they blast out the ashes in a canister and shoot it out the top of the fist over the valley. It goes up about a thousand feet and explodes. The ashes drift down all over here and that's it.'

'That I do not buy,' said Mike. 'We place ashes in different containers, but . . .' he wasn't quite convinced he could make this special explosive container.

'Do you spread ashes over the seven seas, all over?' Hunter enquired. The mortician puffed out his chest and said: 'I happen to be a pilot and sometimes we have the families meet in a particular place and circle over the site and let them pour the ashes out over wherever they wish.'

'That's good!' said Hunter. 'I am a firm believer in the

ashes to ashes and dust to dust concept, though I'm never quite sure whether we feel anything after death — my only concern,' he added. 'What do you do for those who are still living?'

'None of us know how long we are going to live — maybe two years, maybe six months,' counselled the mortician, wisely.

'Doctors tell me I may have two years, six years, two weeks. I may even live another ten years. But that's the way it will be. Then with Bob Dylan's song "Hey, Mr Tambourine Man" playing as it happens. I want to go out on that. That's my funeral! *Res Ipsa Loquitor* . . .'

And that is how Nigel Finch finished his film, adding an animation of the canister being fired into the sky and exploding over Woody Creek.

In spite of Hunter's objections during the making of it, *Fear and Loathing on the Road to Hollywood* became one of his favourite films and he showed it constantly to anyone who visited Owl Farm over the next twenty-eight years. Even to me.

THE FILMING OF
WHERE THE BUFFALO ROAM
1979

Gonzo goes to Hollywood . . . Cut! Print! Perfect! Do it again! . . . A piece for Rolling Stone *magazine*

Where the Buffalo Roam: The Twisted Legend of Dr Hunter S. Thompson opened in cinemas across America. Although the 'legend' referred to in the film's subtitle grew up around articles Hunter originally wrote for *Rolling Stone*, the magazine did not want to be specifically associated with the movie in any way. So, that was more or less that, as far as they were concerned, and they remained neutral to it throughout.

It was in 1977 that I first heard about plans to put Hunter

on the big screen. Naturally I was appalled. *King Kong*, yes. *Godzilla*, well, okay. A remake of *The Thing*, even, or *It Came from a Sewer on the Dark Side of the Moon*. But Doctor Gonzo was really scraping the bottom. Don't listen to me, though. I believe that nothing good was made after *Intolerance*.

I was in LA with Nigel Finch, anyway, making *Fear and Loathing on the Road to Hollywood* and so arranged a meeting with Art Linson, the director of *Where the Buffalo Roam*, who met me in his renovated British Bentley. It was a kind and nervous gesture but this was Hollywood. I was moved into the Universal Sheraton Hotel and Anna and Sadie, our daughter, flew direct from London to join me. When Anna entered the hotel suite the first thing she did was to draw back the curtains to admire the view.

'Oh, look,' she said: 'they've put us above a factory!'

She was referring to the Universal City film lot — a complex of aircraft hangars, sheds and industrial air-conditioning plants overseen by one big black glass tower — the cold heart of the vast Universal Pictures industry.

It quickly dawned on Art that Nigel was here to film the same man and that the BBC hadn't flown me out with a body-guard for the purpose of discussing work I might carry out for the benefit of him and Universal. Here were these two direc-tors, both with one idea in mind. Doctor Gonzo. The only difference was that one man wanted to make a documentary and the other a movie based on fact, but with artistic licence. The atmosphere was not the kind to make the party go with a bang and so the days wore on.

Art liked my drawing, *Spirit of Gonzo*, but it obviously wasn't the one for his movie. Twelve months later I did the poster for *Where the Buffalo Roam* and partly because of it, I suppose, I got a call to record the filming of the movie at Universal City in a set of drawings.

Incidentally, the original poster subsequently disappeared — I presumed stolen by a gang of international art thieves. It was, in fact, stolen by Hunter who was often gripped by an insatiable kleptomania. He stole far more of my work than I realized from the offices of *Rolling Stone*, blaming Jann Wenner whenever something went missing. He did not realize that each time he committed such a felony, he stole a piece of my soul too.

Art's film was to be based on the wild relationship between Hunter and his idealistic Chicano friend, Oscar Acosta. The part of an artist — who briefly appears here and there, draws something and goes away — had been written into the script.

'Is that me?' I inquired anxiously, hoping I was at last about to taste the fruits of international stardom.

'Well, not exactly,' Art mumbled. 'It's . . . well . . . look, Ralph, the story really hinges on Hunter's and Oscar's relationship. It's an integral part of what's been happening in American politics since flower power and up to Watergate. Stuff like that.'

'Well, I'm in there somewhere,' I insisted. 'If the third person is an artist who draws, either it's me or it isn't an artist at all and it's somebody else. It's a perfect cameo part for Robert Redford if it is me, though. Or I could play myself.'

I should have thrown a real artistic fit right there. There have been many times, over the course of my career, that I should have done that, but I was too fucking nice. I kept forgetting that I was swimming with sharks and none of these creatures gave a rat's arse what I might have, or could have, felt.

I am groaning as I write this piece, that I was systematically screwed over any part of this and other projects I was rightfully entitled to through the years. It was a time of thievery and personal ambition and it has lasted until after Hunter's death. I simply did not realize that Hunter's friendship was also a business agreement; he was wise and careful and had surrounded himself with lawyers . . . and guns and other people's money. He was much more into deals than personal affection.

'We'll see,' was Art's hollow reply.

It's difficult to swallow deep rejection when one has been so close to something exhilarating. But I did my best to offer my services in any way that might help Art and Nigel achieve some decent vision of what they each had in mind. Film has a nasty way of nailing the 'truth' to the mast and somehow it stays there like a crucified white shirt until the winds tear its last rotting shreds off the nail. The world passes it by as if it is a mistake and never returns to ask why the nail was ever put there in the first place.

The year before Hunter was living in the Florida Quays, being Hemingway and experimenting with the effects of sudden acceleration on drunkenness by driving his boat clean through a moored yacht and then up on to the jetty.

Later he started discussions with me about the possibility of making a documentary about the film Linson was going to direct for Universal. That one died before it ever began because of the tight union curtain that surrounded all creative activities in the film world then. As Hunter pointed out to me on his first day on location in the desert, twenty miles out of LA, in Piru Valley — a piece of Getty land, so I was told — they should have renamed this film *The Death of Fun*. To be unfair to every one, this remark might have been aimed at the director himself. He had insisted that every scene — nay, every line — be shot at least eight times from one particular angle and then eight times from every other angle, finishing each scene with the classic refrain: 'Cut! Print! Perfect! Do it again!' It must have been hell-on-earth inside his mind, especially at the start.

Art Linson was directing, having taken a four-month crash

course to acquire some of the tricks. I can appreciate his apprehension, nervousness and obsessive caution, and respected them as qualities, but though he's no John Huston, I was prepared to give him the benefit of the doubt. The script wasn't *The Treasure of the Sierra Madre*, either. Nor, for that matter, was Bill Murray Humphrey Bogart, but, Bill, I believed that one day you would be, bless your heart, when the pain of living had begun to score lines across your lovable features; now we know, of course, that he is, indeed, a big star in his own right. Hunter Thompson wasn't Joseph Conrad, Jimmy Carter wasn't Harry Truman. But, strangely, Richard Nixon *was* Richard Nixon.

I'm no Pablo Picasso but there's no harm in straining. After all, the charm of any activity is in the trying and so rarely in the finished article.

Watching Linson at work, it soon became obvious that he was in no frame of mind to catch the abandoned pure essence of Gonzotic madness which can only happen in uncontrolled conditions. But with all those Universal millions plus some of his own, no doubt, lying heavy on his shoulders and enough film to stretch from coast to coast, it must have been difficult to relax. Art's fanaticism for the subject he was trying to portray was undoubtedly there and his sincerity too, but I feel that he was blinkered by the idea that the movie was a runaway hit before he'd even begun to film it, and, therefore, nothing could be left to chance.

My opinion was reinforced when Hunter decided to take a walk around the location to relieve the boredom of being an idle spectator. He absently walked past the dirty window of a shack that was part of a scene which was being filmed at the time. Art's voice, amplified by a bull-horn, thundered out across the valley: 'Cut! Get Hunter out of the movie!' If there was ever a perfect Hitchcock cameo, that was it, and I don't think it even made the cutting room floor.

When confronted by a situation in which nothing creative is happening, one tends to set about on an idea of one's own. This seemed like a perfect time. I had drawn a picture of Hunter as a buffalo, inspired by Art's title for the film and for no apparent reason had named it *Gonzo Guilt*.

It appealed to Hunter, but, being Hunter, he wanted to extend the idea and make it into a set of buttons or badges. Anyone who received one of these buttons would immediately become a member of some secret brotherhood and the object of the exercise was to leak the buttons to various members of the film crew until everyone had one – everyone, that is, except the director. It provided a suitably nerve-racking undercurrent – an alien force off which the director could feed. The plan worked up to a point. Everyone who did not have a button knew they had to have one or face being ostracized.

As far as I know, Art Linson never did get one of the original ones, or any of the other two we made, which read: 'We are not like the others,' and: 'I am a real friend of Hunter S. Thompson.' It was a nasty thing to do and things like that don't make me proud – I wrote at the time – but now I think there should have been more of them. And they do make me chuckle, a mean chuckle. And while I am on the subject of how people behave on and around a film set, somehow someone had learned that they were giving me a four hundred dollar *per diem* expense allowance. I don't know how they knew but people were hanging around me all day, asking for a couple of hundred dollars to tide them over.

The only other event that had any bearing on my involvement in the film was the writing and performing of a song. Hunter occasionally got to know people who had good places to while away an hour or two and, on one of those occasions, I recall that the place Hunter had acquired nestled in Beverly Hills and had a black swimming pool.

I decided to invite Mo Dean, wife of John, over to meet Hunter. As we were a strange mix of very different spirits, it was an extremely creative atmosphere and six hours later we had a song, or the bare words, anyway. It only needed a few more stanzas and I wrote those later.

They didn't use it in the movie and it now gathers dust inside its cassette case on my library shelf. No one realized its potential. God knows how much crap got through the net. Those questions cannot go unanswered. So, for those of you who crave to know what the hell it was that six hours of pure unadulterated creativity produced, here are a few lines. Unfortunately, we cannot hear it, for, after being in existence for almost thirteen years, *Rolling Stone* was still not wired for sound. Now, there is nowhere you can go and not be recorded, and that includes your eyeballs.

Those Weird and Twisted Nights

[The song starts with a quiet romantic stanza which
Mo suggested to soften the Gonzo shriek . . .]

My day is your day
Your day is mine
Our days are no days
Until you are mine . . .

[And then the song proper begins . . .]

Ooooh! Ye-e-es- O-o-o-oooH ye-e-e-ees
Those long strange nights
Those weird and twisted nights
Those weird and twisted nights

A scar heals black in the neon lights
Through weird and twisted nights
Headlights spear approaching cars
Black needles pierce the eyes
OooO – H ye-eee-eee-s OO-ooo—hoo-ye-eee—e-e-s.
Those long, strange nights,
Those weird and twisted nights,

But never mind the nights, my love
Be-ca – a- au -sssse . . .

They never really happen anyway
They never really happened anyway

And days go by
And love goes by
And sooner or later
We all must die
And sooner or later
We all live a lie

Drive your stake through a darkened heart
In a red Mercedes Benz
The blackness hides a speeding tramp
The savage breast pretends

Ooooooo-ooo-oh yesssssssss!
Ooooooo-ooo-oh yesssssssss!
These long, strange nights
These long, strange nights
These long, strange nights
Oooooooo-ooo-oh go-oo-o-od!
Oooooooo-ooo-oh go-oo-o-od!
These warped and twisted nights
These weird and tortured nights

If you write words shocked through with truth
Hunger, dirt and gutter sharp
Eat the words and spurn the gutter
Climb the rise and surf

Ooooooooooooo-ooo-oh yesssssssss!
Ooooooooooooo-ooo-oh yesssssssss!
These long, strange nights
These weird and twisted nights
These warped and civil rights
Oooooooooo-ooo-oh go-oo-o-od!
Oooooooooo-ooo-oh go-oo-o-od!
These warped and twisted nights
These warped and twisted rights
Ah, but never mind your rights, my love
Because . . .
They never really happened anyway
They never really happened anyway

But days go by
And love goes by and sooner or later
We both must die

And sooner or later
We all live a lie

[The song rounds off with the band singing . . .]

They never really happened anyway

[Repeat to saxophone and other band members
And finishing on Hunter's voice saying to me over
the phone . . .]

Well . . . er . . . let me get some whisky, of course.

With Hunter around, Bill Murray was able to study closely
the man he was portraying. I talked to him about my impressions and observations of Hunter, especially his mannerisms –
a combination of a forward swaying lurch, nervous, rearing
twitch and grimacing clumsiness, with a dignified monotone
voice that lent the whole thing a certain nobility. Maybe that's
a bit kind, but you know what I mean. It is to Bill's credit that
within two weeks there were two Hunters on the set. This gave
me an unexpected difficulty so far as drawing was concerned.

I didn't have the same hang-up with Peter Boyle's interpretation of Oscar Acosta. I had only met Oscar a couple of times in San Francisco before he disappeared and my vision of him was not coloured in quite the same, personal way. Consequently, Boyle's strong, wild personality took over easily and his virile performance gave me some marvellous feedback. As did Art Linson, playing the browbeaten first-time director riding his bucking steer at a rodeo show. He became so much part of his own work that it made him very much a part of mine and, thankfully, he enjoyed the result, I believe, vicious as it is.

If I had to recall one sequence that captures the very taste of Hunter's personality, it would be the toilet scene. Bill/Hunter, posing as Harris from the *Washington Post*, in a suit four sizes too small, is shaving when Richard Nixon comes in for a pee. Harris turns as he realizes who has just walked in and, taking off most of his clothes, starts to wash the sweat out of his tennis shoes. Warming to his own performance, Harris launches into a dissertation on the problems with screwheads.

Nixon, who appears to be suffering from cystitis, asks Harris how his family is, and I quote from memory.

'The screwheads have got my daughter, sir,' replies Harris, beating his shoes on the basin as he talks, 'and they'll get yours too, sir. Well maybe not Tricia, but Julie, sir, and you, sir.' The wet tennis shoes hit the tiled lavatory wall with unforgettably dramatic effect. 'They really hate you, sir.'

Nixon winces and leans on the urinal mirror.

'And, sir,' continues Harris, who hits the buttons on two hand-dryers simultaneously and then hangs the shoes on them to dry off. 'Sir . . .'

'Yes, Harris?'

'What about the weak, sir (wap), the silly (wap), the Italians (wap) and, sir . . .'

Harris's voice has reached its most poignant pleading tone.

'What about the doomed?'

The index finger of Nixon's right hand beckons Harris (Murray) towards him. He moves forward, his face a marvellous portrait of comic sadness.

'Harris . . .'

'Yes, sir?'

'Fuck the doomed, Harris!'

It didn't end there. Extremely early one morning in December of 1979 the phone rang. I was still very much asleep when I answered.

'Ralph, it's me, Hunter. Where's that song? How are you? We need it.'

'I'm still working on it with a group called Tax Loss.'

'Tax Loss! Holy Shit! That'll create some attention! Send me a demo. By the way, you've just done a drawing of Bob Dylan, Ralph. It means a lot to me. I can't explain why but the signature on the horizon, *Slow Train Coming*, you wouldn't understand. It's good, Ralph. I need the drawing.'

'I haven't got it.'

'What!'

'I haven't got it.'

'You treacherous swine! Where is it?'

'Still at *Rolling Stone*, I hope.'

'Holy Shit! Jann will have it locked away by now.'

'No, he won't. I specifically asked for it to be sent back because I wanted it myself. He would never break his word with me. I taught him how to be honest.'

'But you don't know Jann. You don't understand real greed. It's like a vulture after meat.'

It was too early in the morning and the light touching the chimneys over the Wandsworth rooftops was too poetic to think about vultures – or early-morning light, come to that.

'I'll get it,' I said.

'When?'

'I'll ring today and ask.'

'No, don't do that. Jann's bound to get suspicious. He'll grab it and you'll never see it again.'

'Not if I want it.'

'Well, okay, if you're sure.'

Two telephone calls later, the drawing was picked up. No one bothered and no one appeared to show signs of suspicion, but the enjoyment of the paranoia, Gonzo style, prevailed. It never fails.

I went back to bed.

'Who was that?' asked Anna.

'Hunter. He wants the Dylan drawing.'

'Oh.'

The phone rang again.

'Ralph, it's me, Hunter. I'm going to Hollywood. No one must know. It's important. I'll be at the Sheraton. Bill Murray is there. We've got to get into the editing suite and change the beginning and the end. The film has no message. It doesn't mean anything. This is an absolute secret. No one, Ralph, must ever know I called. The words of the song – *It never really happened, anyway.* Remember? I never went to Hollywood. I didn't meet Bill. I didn't call you. It's important, Ralph. And we need that song.'

'I'm working on it. It's taking time. I'm not used to the process.'

'Okay.'

He rang off and as far as I know the film is no better or worse than it would have been.

I'd like to think he made the effort, though, without appearing to be involved. It was the least and the most he could do. No jack rabbit just lets himself get run over.

'It is only now as I write this after twenty-five years that I have discovered how the Bob Dylan drawing got into Hunter's kitchen and is there to this day. He went to New York to the offices of *Rolling Stone* and stole it from the editorial department with a lie. The lying, cheating bastard will get no rest. I know where he is!!

And while we are in New York, it was still 1979 and I was there, too, staying at the Gramercy Park Hotel. It was also sometime around that time that the Bob Dylan drawing was done – in a hotel room in New York. I got a call from Hunter with a strange request.

'Er, Ralph! You don't happen to know where I can get some er . . . drugs do you?'

'How the hell should I know where to get you drugs. Surely that's your department.'

'Well,' he said, 'you seem to know a lot of people in this town. I thought maybe one of your pals would know. It's a long shot.'

I thought for a moment and thought of someone who might just know.

'Give me fifteen minutes,' I said, 'and I'll ring you back. Where are you?'

Then I rang an old friend of mine who might just know someone. She rang a friend of a friend of a friend, and within ten minutes she was back on the phone with the information I required. I rang Hunter back and we arranged to meet downstairs and get a cab uptown to somewhere around 67th and 5th. I remember it was opposite Central Park.

Sitting in the cab we were quiet for a while and then Hunter turned to me and said, 'This is extremely noble of you, Ralph.' The gravity with which he delivered his remark made me smile.

We arrived at the apartment block and entered the lobby to be greeted by a concierge. We asked for the floor and took the elevator. It stopped and the doors opened. We were right there in the apartment and before us was a glass table on which lay six lines of coke, all sliced and ready. A small, slim man greeted us and I introduced Hunter.

The apartment was in semi-darkness with a light focused on the table. Hunter was impressed. The dealer invited him to sample his offering and, rolling a dollar bill, Hunter kneeled down at the table level as if he was about to pray. Aiming his flared nostril at the first hit, he sucked up the white dust like an ant eater. Then, he confronted the second line and it disappeared the same way. He dispatched the third and then the fourth with indecent haste, paused, looked up at the expectant dealer who was eager for an opinion, and grunted like a steel hammer.

'Ralph doesn't like drugs, y'know!' he said and then, in a dual motion, nodding his head as he did so, sucked up the last two lines within a second.

'Well, what do you think?' I said nervously, expecting him to rage against the crap he had just ingested.

'Wunnerful!' he grunted, looking up, traces of white powder still displayed around his right nostril.

'Well, which one do you prefer?' I asked.

'The sixth one!' he snapped.

'Oh, come on, Hunter! Of course the sixth one is going to seem best! Are you sure you are not just a little influenced by the build-up. Your judgement must be a little deranged by now, or positively flying!'

'Whatever you say, Ralph.' He rose up, indicating a desire to deal. The little man invited him into the bathroom, the door closed behind them and dark mutterings ensued. I never did find out why the bathroom was the preferred market place but it was not the first time I have witnessed this ceremony.

When they re-emerged, both were smiling, which indicated that a deal had been struck. 'Ralph!' said Hunter.

'Yes?' I responded.

'Er . . . can you lend me three hundred dollars?'

We got back in the lift after polite handshakes all round and the two of us descended.

'Did he give you his business card?' I asked.

'Just remember the street number,' he said. 'I'll need to come back.'

'Okay!' I said. 'But bring your own money!'

THE 80s

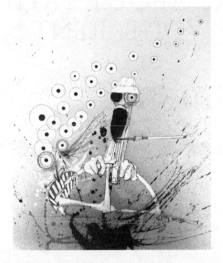

'Ralph, Yr. baroque style of psycho-gibberish is always appreciated here. But what I really need is a six month loan of $50,000; at whatever % rate you can handle. Keep yr. advice & send money. Thanx HST'

'THE EIGHTIES, RALPH, ARE ABOUT PAYING YOUR RENT.'

1980

Running like bastards . . . Partying at the bottom of heartbreak hill . . . Why do they lie to us . . . John Lennon checks out . . . Even the dumbest can make money . . . Lending guns and daughters . . . How to be a raging genius

'The eighties, Ralph, are going to be about paying your rent.'

We were still in the seventies when Hunter figured out this piece of wisdom. In his skewed, logical way, he had thought about it and he turned out to be right.

We set our eighties off in December 1980 with an assignment in Hawaii, covering that year's Honolulu Marathon. We weren't greedy, necessarily — well, I certainly wasn't — but what I did not fully comprehend was that Hunter had developed an unhealthy hunger for anything that generated money. He had sold the rights to *Fear and Loathing in Las Vegas* to several film production companies who knew nothing of each other and equally eager producers who not only wanted to make money themselves, but were desperate to capture the Gonzo legend that was fast becoming a serious item that could no longer be ignored. The bigger the legend grew, the less Hunter required my services, except to embellish the result of his feverish wheeler-dealing that stretched back to our earlier collaborations from the early seventies. I always knew we had forged a lifetime friendship but I was unaware of his avarice or simply thought it was a joke.

It was around that time that Anna and I moved house and I began one of my biggest projects. Looking back over the ten years since 1970 we had still been trying to pay our rent, but there was a hangover from the sixties – a swaying, doomed idealism, twisting slowly in the wind. After *Sigmund Freud* in the late seventies I had decided to take on Leonardo da Vinci, which came to me after I read that Freud had said that Leonardo was 'a man who woke up in the dark'. It sparked an idea about becoming Leonardo, myself, and I began to work on a couple of his projects. It was the reason we bought the big old house that nobody wanted at the time. It is hard to believe now that there was no speculator waiting to buy it for a song and pull it down, giving the wretched little creep five acres in which to cram a new estate of houses and make millions.

It was just before every little guy with a rancid dream of easy riches was taking off to speculate. Even in my local pub, the Chequers, in the mid-eighties, the *Financial Times* was required reading by these would-be millionaires. Only Ben, our local serious drunk, poacher and gravedigger, sneered at the 'snot-gobblers' he despised. They walked into the saloon bar every morning with the *FT* under their arms to discuss the mirage of wealth in a desert of vain hope that it would

be theirs. This included my dear friend, the landlord of the Chequers, Martin Lock, who, within half an hour of Hunter entering the pub, had lent him his gun and his daughter. It was all a dream that was destined to end in tears, but not yet.

One of these people became known as 'Scattercash'. He was a true professional, but was so named because he never once offered to buy a drink for anyone. I have a memory of one of these people trying to sell me cheap gas. He had become a salesman for one of the newly formed companies springing up like Chinese bamboo stems, trying to foist deals onto anyone he thought might have a big house and therefore a big heating bill. I never did accept his offer and I never got involved. The bubble lasted, of course, until Black Wednesday, 16 September 1992, and from that time on you never saw a *Financial Times* in the place. The small investor was finished.

It is a source of pride to me that I have prevented the takeover of my five acres. I am a thorn in the flesh of any cheap pimp who would, given half a chance, rape what is left of this place and bulldoze it into a compound. Fuck those people! That has given me great satisfaction. We are the last bastion against the ravishing of our neighbouring valley, which two thousands years ago was a Roman ragstone quarry, a vine-yard and a natural place of beauty. Hunter referred to the place as my castle in Kent and I did my work in what he called 'the War Room'. It was me versus the world. Cartoon as weapon. Screw you! You are going down the wrong drain! Follow me!

We settled down in the place and from time to time enter-tained on a grand scale. The house demanded it. You can't live in a place like this and not feel like natural aristocracy – the best kind, of course! A friend called David Gollins gave me a goat we called Rebecca. I would walk her down the valley every day and stop off at the Chequers for a pint, hitching Rebecca to the gatepost near the entrance to the public bar. Hunter had his peacocks and I had my goat. Hunter also liked Dobermann pinschers and had two or three of the savage beasts during the late seventies.

He and Laila, his girlfriend, came to stay in 1981. His

visit became a massive intrusion, with notes under the door that read:

```
Ralph,
Do not wake us up for any reason today -
   NO GUNS (No falcons) no shooting, no visits
to Pub - (forget Nigel)
      - No phone calls NOTHING - we will sleep
until we wake up (Forget Aleyo) NO VISITORS
repeat NO VISITORS
      - In a nut, leave me alone until I wake
up & cancel all social plans. We need to sleep
and don't worry about either one of us leaving
the room after we locked ourselves in & we
wouldn't have dared to go anywhere. And I
don't want to talk to anyone today - not
anybody.
   Thanx, HST
```

There were dead of night movements and animal noises, nightly chauffeuring to the pub and a regime of demands that sent Anna into a trauma and serious doses of Valium. One of her fearful anxieties was that he would burn the house down. Our daughter would not take her coat off when Hunter was in the house. 'I can't help it,' I explained pathetically. 'I met this man, by default, and he is my friend.'

He liked to stoke the stove in the library, create a fierce, blazing furnace and stay up until dawn smoking pot and phoning people at ungodly hours.

The eighties were indeed about paying your rent, but in a more pernicious way than we could have imagined. I took Hunter and Laila to London to spread out for a couple of days and introduced them to Bernard Stone and his bookshop. It was a place of refuge for them. They met a monstrous battalion of poets and writers, also in refuge, and hooked up with an 'electronic' company who could transcribe Hunter's notes for *The Great Shark Hunt* onto a new computer system incorporating what became known as 'floppy disks'. They were a miracle at the time. Hunter told me he was showing me the future and it seemed likely. What we did not know back then was that he was showing me merely a glimmer, a morsel of a kilobyte that was to become the wildest dream-reality for millions who pursued this strange new phenomenon. Until the day he died Hunter never took much interest in any part of the electronic bonanza that could have been such a boost to what it was he had in mind for the common man, and for himself.

'We fucked up, Ralph,' I can hear him saying. 'They invented the perfect tool for the New Dumb. They can now flourish in a land of serious stupidity and greed. They can infest the planet with every sick asshole you can dream of and make him sound sane. Now, any cheap, lying fuck can become President of the United States and sound good. Every mindless little screwhead can pour his sickest thoughts into this new machinery, twist it a degree out of normal and send it back out as wisdom. You think we have arrived in the Land of the Living Dead. No, Ralph. We have only just begun!'

He was right of course and the more he raged against the coming of the light as he put it, reversing Dylan Thomas's famous line, the more it made sense to me and to many who found his violent wordplay a tonic for living. When you were searching for words to express your anger and frustration in a world gone wrong, Hunter expressed them and filled others with the answers to their needs.

In 1977, after Nigel Finch finished *Fear and Loathing on the Road to Hollywood*, Hunter and I had individually been doing our own things. His book *The Great Shark Hunt: Volume I*, a collection of his work to date, excluding *Hell's Angels*, was due out in 1979 and, in the same year, a book I had been working on for three years was also due out and my preoccupation with it had made our collaborations less frequent. My book was about the life of Sigmund Freud, taking his 1905 book, *Jokes and Their Relation to the Unconscious*, reconstructing Freud's life through the joking methods he used to explain what a joke was. I took each joking method to draw and write my way through his life. What had started as a book about Jewish jokes became for me something far more interesting. I went with Anna to visit the Freud Museum in the 9th district of Vienna and spoke at length with Hans Lobner, the curator at the time. He allowed me to lurk in the basement which was Freud's first consulting room, looking for clues, picking up old bottles, replacing them exactly where they had been, trying not to disturb the karma, feeling the vibes and generally doing what most normal people do not do, like lying down in the very spot where Freud's consulting couch had been. That was a strange experience because I was now certain that I was under the spell of this strange man's thesis on the condition of the human mind. But it was the only way to go. In fact, it is the only way I go when I attempt to do anything creative. I must make contact somewhere with the source of where the specifics happened. To me it makes such vital sense.

As Hunter expressed it: 'It's Ralph's gibberish about the gibberish of Sigmund Freud' — his only comment about my book that I can recall. He never commented on the drawings,

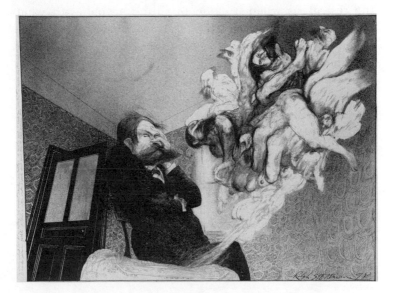

which I consider among my best in any book. It was something so weirdly foreign to him that I may as well have done a book about the history of the cake, and while I am writing I am saying to myself: 'Holy God! That's not a bad idea!!' Cakes, which include poison cakes, bomb cakes and drug cakes, is a subject to be taken up at some later date.

In the eighties, after *The Curse of Lono*, Hunter became more circumspect about my involvement in anything to do with Gonzo, as though the very presence of one of my drawings in a journalistic project of his own represented a serious threat to his domination over the world we had collectively created a decade earlier. My drawings were becoming baggage, best dropped off in some bushy scrub along the trail, halfway across a wilderness, or in a dirty pond along with old bicycle frames and rubber tyres. Writers are like that. Whether they like it or not, whether they attempt to consider themselves actual members of the human race, or chosen spokesmen for life's underprivileged, winners of prizes or rich and curious seats of learning, I had, as far as he was concerned, exhausted my usefulness. But in his moments of quiet loneliness, I was still there as an integral part of the Gonzo spirit. The poor bastard was as alone as the rest of us when it came to filling a void with what most of us believe the creative spirit to be. These are mere speculations, but even as I write now, in my own chosen loneliness, missing the man like a lost leg, I realize our collaboration was one of those venal necessities I cannot brush aside, and neither could he.

The rent! Ah, yes! the rent. Hunter and Laila coming to stay, weird shenanigans at large in a foreign land, brought us to the finishing of the strangest of our collaborative efforts, *The Curse of Lono*. Throughout the creation of the book, its text and its drawings, he needed the drawings before he could attempt to finish his part of it, and he always referred to the book as 'your next book' — not his. He was trying to sever our ties but this book had to be done in spite of his serious objections to the idea, and it stuck to him like a horsefly on shit.

To Alan Rinzler Bantam books
Dear Alan,

Let me be the first to say it, old sport:
You may turn out to be a credit for yr. Race,
after all . . . or even to Our Race, which is
probably not much different in the end . . .
or The Beginning, either, for that matter
. . . But these evil times somehow conspire
to make us all guilty, for one reason or another
. . . and, as always, for good or ill.

Which brings us with a fast & terrible
suddenness to yr. Letter of Oct 6, regarding
the possibility of a Bantam Book by me and
Ralph for a massive advance & a tall & permanent headstone in the history of whatever kind
of 'literature' we've been locked into for
lo, these many years.

[. . .]

So let's do it.

My plans at the moment are to spend the
month of December with Ralph in Hawaii. I've
already rented a compound with two houses and
a pool on the Kona Coast Beach: One house for
me & Laila, and the other for Ralph & Anna
& Sadie . . . we also have a boat, a jeep &
a personal tavern a few miles down the road.

You should make plans at once to come to
Hawaii and work on the book with us. We have
to cover the Honolulu marathon for RUNNING
during the first few days in mid-Dec [1980].
But the rest of the time I've set aside for
working on the book with Ralph . . . which
is why yr. 10/6 letter came as such a fitting
& pleasant surprise. It sounds on its face
like the same kind of hopeless bullshit that
Sam Metnik wanted . . . but Sam Metnik did
not put together Ralph's 'Amerika' and 'F&L
on the Campaign Trail'.

So I'm taking you seriously, Alan – and since we were planning to do a book in Hawaii, anyway, yr. timing was almost eerie (if only because yr. letter arrived just in time for me to use it in court as what turned out to be convincing evidence that my request for a 'continuance' in the never-ending divorce case was not because I wanted to take 'a vacation in Hawaii', but because I genuinely needed the time to WORK FOR MONEY – which I do, but it would have been a hell of a lot easier to get the continuance if I'd planned to spend December in Pittsburgh, instead of Hawaii . . . so now I need a bit of tangible evidence, like a book contract, if only to maintain my credibility with the judge. The final court gig is scheduled for Feb 9, 10, 11 & 12.

   [. . .]

   Meanwhile – if we're talking about a Real & Serious book – I think an advance of $125K (to me) seems about right, given the sales figures for all of my books and the fact that I've never written ANYTHING that didn't make money for somebody. And usually for everybody.

   I figure $100K for whatever I write, plus $25K for editing and organizing the whole thing – because I will insist on Final Cut on any book that has my name on it, as author.

   [. . .]

   Can you pay $50K on signing, $50K on delivery and $25K for whatever editing and organizing work I'll have to do all the way to the Final Cut?

   So let's keep it simple . . . and call it something like 'Fear & Loathing in THE KINGDOM OF THE BLIND'. That would give us room to kick ass in all directions – but especially in the area of connecting the HST/Steadman wisdom of the 1970's & even the '60's to survival in the '80's.

[handwritten note:] Ralph will probably want money. But so what?

Will history eventually judge us as fools and charlatans? Can we bow our wretched necks & take the Kona Coast in 1980 like we took Washington in the '70's & the Kentucky Derby in 1969?

Or Zaire in 1973?

I think it's a valid question, with terminal implications for a hell of a lot of people beyond you & me & Ralph.

If we can create a high mix of Memoirs & Madness that is also a Milestone with even a hint of continuity into this weird new world of the '80's, I think we can lash out a classic that will sell at least a million copies, if it's properly managed. We are looking at a whole generation of refugees out there who desperately need a book (or a film or a statement in some form) to give them a sense of personal continuity in a world they never expected to have to cope with.

It is a long, long way from Oscar's world to the one we live in now – and if Oscar failed to make that leap, he had plenty of company.

Maybe it was impossible. Or maybe it was all a dumb, wasted trip in the first place.

But it was a trip, and anybody who missed it has a right to feel cheated, for reasons they'll never really understand . . . Like a lot of the people who didn't miss it can't understand why they feel so cheated now.

Yeah. We have a problem, Alan – and just thinking about it has made me too wiggy to type more than a few more words.

Call me fast on this one. We're going to
Hawaii anyway & and our only justification for
renting the compound for a month at $4000 plus
exp. is the Knowledge that we will get a book
out of it . . . And we will, Alan; for you
or somebody else . . . and we shall kick ass
on the Kona Coast . . . Because we must.

Right?

Yes. We will skulk off the plane in Honolulu
with the hopes & dreams of a whole generation in our hands. I am already entered in
the Marathon & I plan to enter Ralph in the
Pipeline Masters surfing competition . . . we
will not win these weird events, but we didn't
win the Kentucky Derby either.

We did, however, PREVAIL. Which is as
different from winning as a forest is different
from the trees.

WHAT?

Did I say that?

Yes.

Which means I have to go now, Alan. The
full moon is making me strange.
So . . . let's jump on this bugger ASAP &
have it to the printer by June. I sense an
Opening here, but it interests me.
We can call it EAT SHIT & DIE.

Why not?

HST

HAWAII
1980

Why run? . . . Weeping on the black rocks of Kona . . .
Hunter's firework display . . . Old buddies at Huggo's
bar . . . Thieves and peddlars . . . God's own drum roll

The *Lono* adventure had begun in 1980. It was Christmas and
I was in Hawaii for another weird assignment, on the strength
of which Hunter and I had agreed to rent adjoining beach
houses on the Kona coast for six weeks. He had brought Laila
and I had Anna and our daughter, Sadie, with me. The
weather was stormy, the Pacific swell threatening to over-
whelm the black rocks opposite our compound. We had originally been commissioned by *Running* magazine to cover the
Honolulu Marathon and this had led to a bigger project, the
book *The Curse of Lono*.

The main thrust of the *Running* article was 'Why do

these buggers run?' Were they running for some serious reason, like for their lives? Eight thousand apparently sane people would be pitting themselves against each other over twenty-six consecutive five-minute miles and only about six had a chance in hell of winning. Most would be coming from thousands of miles away and would be paying their own way. It was and is an insane way to keep fit, particularly in a bug-infested, humid, jungle atmosphere, carbo-loading on beer and spaghetti. Half of the contestants were short-legged Japanese and they would be running from and arriving back within a stone's throw of the wreck of the USS *Arizona* Memorial in Pearl Harbor, on the very morning of 7 December 1980, thirty-nine years to the day after America was shaken to the core by the unprovoked attack that brought America into World War II. Were the Japanese secretly gloating or was this the twisted respect or homage paid by a generation of runners whose fathers would have been the very men who piloted the planes that blasted the hell out of the US Navy.

But the trip became bigger than both of us as we wept on the Kona coast, wondering what in hell we were doing there if it wasn't to create a story, and, worse still, wondering why

was everybody lying to us? It became our daily mantra and we pinned the phrase up in both our kitchens.

It was Christmas Eve and we had planned a party around the pool. Living with Hunter in close proximity had its hazardous moments. Halfway through the night he would stumble into our living room to borrow our TV set because his wasn't working. Sadie was sleeping in a sleeping bag on the floor near the front door, because of the possibility of the Pacific getting rougher and trapping us in a back room with no escape. If you can imagine Hunter galumphing in the gloom to make a beeline for the set, lifting his feet like a paraplegic with alarming and twisted movements, then you can imagine him swivelling precariously to avoid bringing his foot down on Sadie's head. She was blissfully deep in sleep, probably dreaming of the little boy called Zeddy who lived across the yard in a house on stilts, the front of the house having no wall. It formed part of the complex we were in. We were dangerously close to the Pacific side and could feel the pounding spray clouds hit the rocks which provided our beach frontage. Sadie thought the world of Zeddy and followed him around every day like a pet dog. By now, Juan, Hunter's son, had also arrived for the festivities.

I would work at my drawings during the day, and meet Hunter later with his new Kona buddies at Huggo's Bar. Many of them were either real estate agents, drug dealers, captains of fishing boats or maybe all three. Real estate was becoming big business and every time Pele, goddess of fire, had another fit and spewed out millions of tons of lava, these new-style real estate dudes would buy it as new land, as soon as it had cooled, and put their markers down for later exploitation.

Shady people of the shadiest kind fascinated Hunter and, like a fly detecting a heap of shit, he would alight, to the manner born, on such a dump. 'All part of the job, Ralph. You will never learn anything, or stay ahead, unless you mingle, otherwise the dump's on you.'

He was right, of course. Except that all of them, to a man, loved football and that was the most off-putting aspect of their wretched lives. Wherever men mingled around Hunter, which was usually in a bar with a TV, or watching it in his kitchen at Owl Farm from his favourite chair, at his desk on which sat his Selectric typewriter, these wanabees grunted and whooped and sucked on beer cans trying to sound like Hunter. I found this the grimmest part of the entire charade. It irked me with a passion and I preferred to spend my time with his peacocks. Hunter knew this and left me to my own devices, which on more than one occasion found me dressing as a woman and embarrassing the hell out of him as his 'Limey friend'. 'You never can tell,' he would say, 'Limeys appear normal but the weirdness runs deep.'

On Christmas Eve we played party games and dressed in Hawaiian *leis*, the flower necklaces visitors are presented with when they arrive on the islands with the greeting, 'Aloha!' Hunter said he was working on a special event around the pool which involved fireworks and special effects — just for Sadie. First, though, we played party games. The most memorable was 'The Farmer's in His Den'. Each person holds hands in a circle and walks around the 'pig' in the middle. Sadie was the first in the middle and we revolved around her. Hunter, who believed that all children were small grown-ups,

made a grab for her and swung her around his head like a doll. She was petrified, until we got her to understand that Uncle Hunter was merely doing what they do on all proper farms when they play 'The Farmer's in His Den'. So she had been given the proper grown-up swirl. There were tears, but Laila and Anna comforted her.

Then came the fireworks display. Hunter was ready and we took our places as far from the pool as possible. This was Hunter at his best, organizing a dangerous but dazzling display of pyrotechnics that could go anywhere but in the right direction. Somewhere in town he had found a place to buy the Chinese firecrackers with which Hawaiians celebrate New Year. There were no restrictions on the strength of them and each firework was packed with gunpowder and claimed to have over four thousand bangs in each bundle.

The first part of the show was pretty enough − Roman candles and shooting stars over the Pacific. Hunter had a pronounced aesthetic side, but no one was ready for the machine-gun rattle of four thousand hellish bangs that Hunter described as 'God's Own Drumroll'.

Each bang was followed by a blood-red shower of the paper that was used to wrap the explosives. The smell of cordite was pungent and hung around for days. When the display was over, the immediate area for about half-a-mile was covered in a red, smoking glow of burning bits of red paper that looked more like a Martian landscape than the black rock coast of Kona. I detected in Hunter, in any of these ball-breaking displays of boyish high jinks, a pride in what he had done. He was still the 'not-so-little boy' from his childhood days in Louisville, Kentucky, the desperate days of a sensitive soul outlawed by the general normality of the decadent viciousness of Kentucky high society, because way back then, all he wanted was to belong.

It was a mere step from that feeling of rejection to an all-out scream into the blackness of a wounded creature that had no identity but its own.

Over the six weeks Hunter and I had developed a strong relationship with a local boat owner known as Captain Steve, who later became a significant character in *The Curse of Lono*. We didn't know his second name. He took us out on fishing trips which were completely abortive because we caught no fish. It was not until Hunter returned to Kona on a third trip that he made his magnificent catch.

Captain Steve's father lived in a condominium up the road. Somehow Hunter had discovered that he was playing around with somebody else's old lady. Hunter came to the rescue like an outraged missionary with an old-fashioned conscience. He still had one of those filthy firework bombs and, knowing where Captain Steve's father was holing up with his little piece of dolly, he set fire to this pyrotechnic monster, right in the front doorway of the father's condominium. With hardly time for him to pitch himself into the undergrowth, it went off and blew the door open on the amorous old man's love nest. Is that a very moral thing to do or what? The man could have had a heart attack. Instead, he came outside in a flannel nightshirt with such a look of terror on his face that it filled me with sympathy. I couldn't stop laughing. Sometimes I asked myself these moral questions and then rejected them as a silly sentimental urge that rationalized my stupid sense of decency. I had become a part of the reasoning that Hunter used on a daily basis to get his own back on an amoral world. It all made such terrible sense.

I began to assure myself that if Hunter was wrong, one day he would get his own retribution. No bastard gets away with it for ever.

Back in England we had decided to go fifty-fifty on the *Lono* project. I had suffered enough! That is what happened. He couldn't work without the drawings, so I did them first and the bundle was sent over and mounted on the wall of Hunter's 'War Room' in order to fire him up. Laila did her best to ensure that this happened and then watched over the tortuous process of trying to make the story work. I had already warned her that these drawings were precious and must be returned after use. It is to her credit that she made damn sure that they were returned.

We had been corresponding regularly about the plot and

its possibilities. Part of the structure of *Lono* was made up of letters to me about the crippling back injury I received when I fell onto a white coral reef. A chiropractor futzed with the muscles around my spine during many sessions, but it did the trick. Ten years ago, I watched Hunter receive the same type of back manipulation from a physiotherapist in his own kitchen in Woody Creek, when he could hardly get up and stagger over to his control chair to confront his typewriter.

I had begun the Leonardo project that was to take up so much of my time. It became both a saga and a refuge because, while considering Leonardo's notebooks and drawings, I had been feeling for a wretched length of time that nothing was going to work on a conventional level. Something was missing and it was fundamental to the making of this story and book of drawings, no matter how much I regarded Leonardo. I had bought his notebooks at the age of fifteen in a WH Smith shop in the North Wales seaside town of Rhyl on the bargain basement table for five old shillings. I had kept them, delving into them over the years, and they had formed the basis of what became my library; small at the time but precious. And that's when the idea hit me as I scanned them again after all

those years. I knew so much about the man, he was practically part of my childhood. Other books by proper experts
had said all that could be said about him, so what was I doing
trying to do the same thing? It wasn't until two-thirds of the
way through my toil that an idea struck me that suddenly
seemed so obvious. Why didn't I try to *be* Leonardo? *I,
Leonardo* became the title. If I wrote in the first person I
would surely know what he did and how he felt.

It worked for me and still does. I painted *The Last Supper*
on our large bedroom wall and it is still there. I constructed
a flying machine out of bamboo and an old nylon tent. Every
three months, over the next three years, a Channel 4 TV crew
came down and filmed the book's development.

These were the golden years of documentary arts
programmes. At the time I was painting *The Last Supper* we
were having the house painted by a wonderful couple of
partner painters called Lou and Pete. Michael Dibb, the
director of the documentary, miraculously interwove the two
activities as we all went about our respective art forms. Me on
the bedroom wall and Lou and Pete on the window frames.
As for Leonardo's flying machine, we took it onto the flat roof
behind the pitched roof of the main house and Lou, who
always said that he fancied himself as John Wayne, wanted to
hurl himself into space on my contraption. I let him hurl the
machine from a pair of steps, which added height to our enterprise, but insisted that he let go of it while I held him by the
seat of his pants to make sure he did as he was told. The
machine glided gracefully out of his hands and, as chance
would have it, got snagged on a telegraph wire, and stalled
backwards, projecting a poetic glow of sunlight through the
bamboo structure and onto the nylon fabric, which was caught
on film. An attempt was made to launch a human-sized
Raggedy Ann doll but nothing topped Lou's accidental play
on light. The film was called *Don't Tell Leonardo*.

I have no idea what was I doing in Washington on 31 March
1981, but I had produced two drawings and I sent copies of
them to Hunter and Laila. One was *The Babel Tower and the*

Attempted Assassination of Ronald Reagan, and the other was a desultory picture of Nancy Reagan as Marie-Antoinette saying to Ronald as Louis XVI: 'Let them eat jelly beans!'

I wrote to Hunter on 30 March 1981:

> I worked like a bastard to pay for my hotel room in these last few days. The New York expense, shuttles and fares from London added up to a sizeable up-front commitment - so thank god for my prodigious talent! We'll get the bugger back with blood to spare. For every time there is a season and it's time to start calling it in.
>
> The contract monies are, as far as I have heard, in the bank and all is green for go. I have begun the big picture for Washington - The Babel Tower - where so many languages are spoken - no one can communicate so the tower is never finished and, most important of all, it is a symbol of man's futile attempt to reach heaven. The other is Nancy Reagan as Marie Antoinette and the 'let them eat cake' legend. The road to the guillotine and a Tale of Two Cities.
>
> Anna just casually came into the room and said: 'By the way, there's been an assassination attempt on Reagan.' Mother of God! I think we got our hotel reservations wrong again - I just saw the news flash - and I do believe it is like recurring flashbacks or, better - a video loop. Relentless, predictable, insane and mechanical. You had better elect a bullet-proof robot next time whose computerized mind contains the spectrum of a collective consciousness. Maybe you don't really want a president at all. Just French cannon fodder - chuck him over the top like a clay pigeon.
>
> Hey ho! - it makes huge changes - maybe

one can only confront this whole subject from
some remote place like Hawaii as though one
were constructing an Odyssey of mythical
proportions not of our world. More of a Greek
Tragedy was my first reaction and then I
thought of the Law of the Splintered Oar. On
that basis Reagan will have to pardon John
Hinckley (Reagan's would-be assassin from
Evergreen, Colorado). Maybe Reagan will find
wisdom and seek a new way as did King
Kahmehameha the 1st.

While we were on Kona we heard of the City of Refuge.
According to legend, the Law of the Splintered Oar, proclaimed
by King Kahmehameha, decrees that if you are being hounded,
perhaps for committing murder, but can make it to the City
and stay inside its walls for a week and survive, then you are
free to leave; you are pardoned and allowed to go on your way.

The main building was a grass hut surrounded by
grimacing effigies. It became the setting for the closing chapter
of *The Curse of Lono*, where Hunter decided it would be the
place he had finally found to write. He would be guarded by
a ranger, an angel spirit, who would bring him whisky, drugs,
women and food every night. I think it became his favourite
tale of the unexpected – his nemesis as a writer. It was damn
near a 'proper story'!

This is all very well and on Sunday I put
the London Marathon on video and watched it
in slow motion - it takes 26 hours! I enclose
some news cuttings to show you how thrilled
everyone was here. It was a monstrous success.
You should have been here but let it pass now.
As I've got the video, we'll watch it someday.

The Telegraph have finally turned down our
Running piece. The letter is obtuse (I enclose
a copy), and rather odd - but it still says
no. The way the marathon took off in a rather

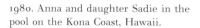

1980. Anna and daughter Sadie in the pool on the Kona Coast, Hawaii.

Anna and Sadie pose on the diving board of the pool overlooking the Pacific.

Dramatic Kona coastline.

Laila relaxes by the pool at the compound.

Sadie dressed for the Christmas party, December 1980.

Ralph serenades on the verandah of the compound.

1980. Hunter reflects at the edge of the City of Refuge on the island of Kona, Hawaii.

Totem effigies protecting the City of Refuge from the outside world.

Ralph as 'buggering fool' outside the City of Refuge.

Ralph taking pictures around City of Refuge.

Makeshift drawing board and anti-splash screen at the Kona compound where drawings were conceived for the book *The Curse of Lono.*

City of Refuge
Hunter

Dear Ralph,
We Killed
like champions
(Lono)

HUMDINGER

1981. Hunter catches his fish on his
third trip to Kona.

Hunter sets the rod.

The catch on board.

1994. Anna makes friends with Hunter's stuffed wolverine in the back of the Red Shark.

Nocturnal visit in Aspen rental house. Hunter attacks Anna with his toy hammer.

Ralph and Hunter in limousine on the way to Juan's wedding, in the hills outside Boulder, Colorado.

Hunter adjusts his son's bow tie prior to the wedding.

1994. Lawrence, Kansas. Ralph
showing William Burroughs his art.

All pals together: William, Anna
and Ralph.

Ralph shooting and William shooting.

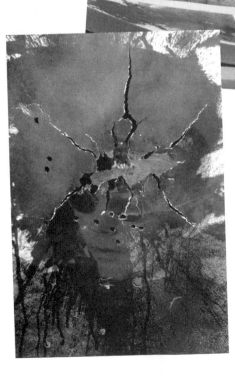

1994. Gun-blasted ink on paper used to make stencil for Sheriff print laid out on the trunk of the Red Shark.

1995. Joe Petro watches Hunter sign the limited edition of the Sheriff print.

Sheriff shotgun art. Detail of Sheriff print showing blast.

Hunter's weapons arsenal, War Rooms, Owl Farm.

1996. Hunter places the blow-up doll on his tractor prior to shooting a propane gas cylinder 'bomb'.

The bomb explodes.

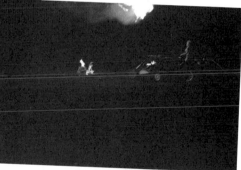

Ralph and Hunter getting ready to shoot ink bottles at artwork.

1994. Hunter visits the Steadmans at their rented house in Aspen and offers counsel.

1996. Owl Farm. Hunter gives Ralph, dressed as a woman, a rare embrace.

Ralph in performance — drag act.

Hunter signs Flying Dog Ale poster.

Hunter uses electric-shock gadget to warn off excessive video-filming.

English way (not a black man amongst them!) has somehow overtaken the Honolulu one unless some bright editor has a better way to present it. Let it reemerge in our book.

God bless. Ralph. PS. I recall owing you $200! Was that a fixed bid?

PPS You no doubt realize that it will be even more difficult to criticize Reagan and his Administration - a weird wave of support will flow over the man now and for at least 12 months he will be bathed in the light of a Saviour - on the 3rd day he rose again - the born-again President.

There was Bush miles from anywhere - Haig on the scene and ready to snatch power, pull the trigger, the military move in and Wap! You have a military dictator - but the dumb kid was a bad shot. What a sick place.

There is an Edgar Allan Poe story called Imp of the Perverse where a young man decides to act upon anything that enters his mind. At first his transgressions are minor ones and then inevitably he thinks of murder and, of course, commits it. Then it enters his mind to confess, not from any sense of guilt, but just because he thought of it - so he confesses - and nobody will believe him - he tries and tries again but to no avail. He finally turns into an alcoholic lunatic - not because he thought of it, but because he becomes obsessed with confession and the frustration of ridicule - because nobody believes him.

Now of course murders are committed or attempted on camera - snuff movies for International audiences - instant notoriety - rock star fame in a split second - it's all so easy - utterly futile and pointless. It's the aspiration of despair. OK Ralph.

THE ELUSIVE BLOATER
1981

*A tale of blundering courage at sea . . . Patience in the
fighting chair . . . Lame jokes and sick thoughts . . .
Notes from a cold climate . . . The City of Refuge*

I returned to Kona with Hunter in May to lash together the
basis of a story that would become *The Curse of Lono* and
during this time I kept a diary.

Kona, Hawaii: Monday 23 May, 1981
    I had been in Kona again for about a week and was just
beginning to get used to looking at boats, stepping on them,
standing on them and acclimatizing my stomach to the possibility of staying on one for hours – maybe days.
    I had been here the previous December, to cover the
Honolulu Marathon for *Running* magazine, which had nothing
to do with presidential campaigns or relentless incontinence,
but everything to do with guys and girls pounding the tarmac
in cushioned, thermodynamic running shoes, collapsing at the
seventeen-mile mark, at the bottom of Heartbreak Hill, in
heaps of sweating blubber and sobbing like babies. John
Wilbur, an ex-football player, had a house at that very spot
and had thrown a party for those of us who were just there
to watch the agony and the ecstasy.
    We held glasses aloft and cheered those who were still
going strong, jeering those who were beginning to flag and
stumble. 'Run, you bastards, run!' was too much of an insult
for some, who would respond angrily with 'You sick fuckers!
How can you stand there and do that? There's not an ounce
of sporting decency in any one of you!', which finished them
off right there, of course. Anger would consume their last

dregs of energy and starting off again was simply out of the question. We would mock them mercilessly as they dragged themselves along until a pick-up bus came by to rescue them, poor wretches of broken dreams. The response of the runners was shameless. We were reduced to collapsing heaps of drunken laughter.

The plan was that Hunter and I would come off the starting chocks at a ball-breaking pace and keep running until we were out of sight around a bend. Then we were to be picked up by a waiting car and taken swiftly to John Wilbur's party. Unfortunately, a couple of days earlier a Pacific wave had lifted me off the ground and dropped me on my back onto a coral reef. I was out of action for most of the rest of our stay in Honolulu. I remember it so well especially as it was the week that John Lennon was shot. I was unravelled by a chiropractor and the following week we were off to spend Christmas in a compound on the black rocks of Kona. Things began to look good for a while . . . but, to return to May 1981 . . .

Standing on the pier the night before, I'd watched the big fish come in. Four- and five-hundred-pound marlins winched ashore with monotonous regularity, the fishermen photographed with their catches like minor celebs, hula-girl groupies hanging on their golden, hairy arms. So, I knew it was possible. Somebody was catching them and a lot of those fishermen hoisting their catches up off the decks to be weighed looked more like sunburned golf pros, with even more golden walrus moustaches and a hobby.

'Listen, Steve,' I said to the captain of a boat called the *Haere Maru* that very evening. 'The way you've been acting these last few days it's obvious you need a change of tactics.' He nodded passively.

'Fresh blood perhaps?'

'Stinkier bait,' I replied, 'or raccoons. Something different and interesting. Fish are like people. They love anything new. What about ahi?' I said. 'Fish are cannibals. Just like estate agents, they eat their own.'

Normally nothing would get me near a boat; nothing short of a press gang, that is, and here I was, getting dangerously close to offering my services. And not just as a low deck-hand help, either. The very finest, no less — first-chair fisherman.

'I've fished most of the best rivers, Steve. Ireland, Scotland, Ticino in Italy, the whole of France. Tactics are everything. Brute strength counts for nothing. Macho won't catch fish, Steve. You should remember that.'

Captain Steve nodded again. His eyes had, by this time, become permanently grafted on to his sun-shades from nights — possibly weeks — without sleep and the nagging torment that a man amongst men could become a figure of total ridicule.

The wreckage of his face hung about his sun-shades in festoons of fleshy blubber, like the innards of a strung-up ahi fish bulging from a gaff wound. Tomorrow was the important day of the tournament and any straw was worth clutching at. Captain Steve wasn't likely to survive much more pain from failure. Mitchell, his chair-man had left that day, but had convinced me the night before that I was the man to take his place. More than that, however — talking boastfully at the Kona Inn Bar, known as Huggo's, about fine fishing in freshwater rivers, we had become blood brothers.

Trout-fishing had been my speciality. I remember catching

one as a boy; it had been about six inches long and I had never forgotten the excitement. That first sudden tremble on the line as it took the bait – dry-fly as I recall. Since that day I had always considered myself a fisherman of the highest order but I was more concerned with the art of the sport. Dynamite wasn't for me. I was a member of the elite.

As years went by, that experience had grown in my mind. When we, as a family, started going to France and Italy I set myself up with a fine rod and tackle and I fished. Always, I fantasized, the next day would be the big one. That day had not yet come, but I persevered and bought more tackle. I had convinced myself that I was there for the love and the antici-pation, swept along by the sheer pleasure of a fast-moving river. I was still waiting. The pure experience. In those thirty-odd years the thrill had taken on legendary proportions. So what was the difference?

Sitting on the chair on board a small boat, trawling the high seas off the coast of Hawaii or casting wide on a river in the mountains of Ticino or from the bank of a potent little river in Llanfairtalhairn in North Wales, where my first fish came to me – what would Hunter ever know about the thrill of that?

My specific problem, however, was overcoming my fear of boats and there I was trying to do just that by convincing Captain Steve that art was all that mattered and bait was what was wrong until now.

Whether the rest of the crew had been listening in to my drunken pep-talk it seemed to be clear by the time we all left the bar that evening that I was in and tomorrow was prepar-ation day.

Sunday, 24 May

Even Alex, Captain Steve's second chair-man seemed resigned to the fact that I was to take the fighting chair. And anyway, after mounting disappointments, they were a spent force and I had come halfway around the world to get the job done.

Hunter stepped on board, eager to gaff. 'No more fuck-ups!' he declared. 'This is the one! Let's kick ass!' Maybe I was the missing link they had been waiting for. I was convinced that the 1,500-pounder we needed to win the tournament was out there. Today was a rehearsal.

Captain Steve said he had slept but he still looked like a haunted gargoyle.

'Don't worry Steve – glad you got some sleep. You needed it. And I simply want to know my Captain is in charge – more than that – in control.'

'Sure as hell will be,' he slurred. 'No problem – we'll catch a big one tomorrow – I promise you that. You guys won't be disappointed. Aw hell, you know me. We'll be out there with the best of them. W . . . slur . . . y . . . slur.' His face was, speaking as an artist, still a war-torn landscape of dribbling spittle.

'Let's do a test run!' Hunter was eager just to get on the water. He was going out there to be Ernest Hemingway in *The Old Man and the Sea*. I was going out there to test my stomach. We both had our priorities. The water was calm but my stomach told me otherwise.

A strange quiet falls over people when they are surrounded by the sea. Everyone wishes he was somewhere else; anywhere but here. Maybe it is the human psyche's defence mechanism against its worst thoughts. Cannibalism, to name but one. Steve was up on the poop deck, scanning the horizon, coming to grips with his fatigue. Chipper was trailing his hand in the water and smoking a joint. Buddy had opened a beer and was standing fiercely at the prow.

I sat queasily in the fighting chair, trying to get a feel for what it might be like to hook a big one. 'Can I put a line out just to play?'

We had had lunch and there was salami, smoked salmon and leftovers. I insisted on attaching some smoked salmon to the spinner to demonstrate my theory of old fishing techniques, but we trawled for another hour without a nibble on the line. I reeled in and tried a slice of sun-dried salami. I was certain that the older the bait, the better.

Nothing.

It was an uneventful day, and for all Hunter's kick-ass rallying cries, when fish won't take the bait, there's not a lot you can do. The rest of the crew watched the horizon and seemed to do nothing, clothed, I thought, in their own thoughts. I took a few pictures of Hunter reading the local rag, *The Hawaiian Gazette*. He seemed very preoccupied. I gazed into the water and willed a fish to come to me. By now, my seasickness had subsided into a slightly euphoric light-headedness brought on by a few gulps from a bottle of Wild Turkey and a beer.

Captain Steve came down the steps and suggested we should call it a day and maybe tomorrow move south where he knew there were plenty of fish. 'Why do they lie to us? There are no fish,' crossed my mind as I reeled in. Meanwhile, Steve steered a course for the harbour.

It had been pathetic and futile but tomorrow would be different and I was getting used to the swell of the sea. That night, Huggo's Bar was heaving with contestants and all the talk was of tomorrow and the big tournament.

'There are no fish,' I said many times that evening, which became 'Thesh no fish' by the time I got up and staggered off to get some sleep.

Monday, 25 May

I woke at around 6 a.m. and immediately took a shower, feeling my nervousness buzzing through a faint hangover. I longed for the day to be over.

Hunter called – brightly, I thought, for him. 'Oh you're up. Good. I'll be around in about half an hour. We'll get a few groceries, beers and stuff at the Union Jack Stores.'

'Okay. I'll be in the coffee shop,' I said, 'having a spot of breakfast. Collect my thoughts. At least it's a calm day!' I added hopefully.

'It can change, Ralph . . . it can change,' he replied ominously before putting the phone down.

We arrived at the boat a little late and found the rest of the crew, Chipper, Buddy and Steve, fuelling up in the harbour. Alex was missing. I climbed on board just as money was needed to pay for the fuel. I did wonder about that at the time. Why always me? I offered twenty dollars and the bill was around thirty-nine. Someone else coughed up the rest and we took off to make the starting line by 7.45 a.m.

Incidentally, it wasn't the *Haere Maru*! Oh, no! It was a fast, outboard, light boat more suited to joy-riding than serious fishing, but it was all we had. Captain Steve's boat had been laid up the night before owing to a series of inexplicable decisions by him which culminated in a disastrous last-ditch attempt to practically rebuild the engine on the eve of the Kona Gold Jackpot Tournament.

The previous night I had been confronted by wires, engine parts and human legs, smeared with grease and sweat, protruding from every hole in the bowels of the ship. But I didn't tell you about that because it would have broken your heart. I turned it all into a drunken haze and tried to shut it out.

I couldn't tell if the engine was actually in place, out of place, tilting for a special reason, or if this was the boat at all! It was just a dim shape, lit only by a mechanic's inspection lamp, up on chocks and too far from the water to convince me that superhuman effort was all that was needed to float it.

'That's our boat up there,' Hunter grimly pointed out. 'Steve had some insane impulse to change his engine and so far no one has been able to explain the wisdom of it. We're doomed, Ralph.'

I was not unduly perturbed – just numb. All this way for yet another disaster – another brutal kick in the nuts. So what's wrong with a few more? It's what life is all about. How boring if it were all to work like clockwork. We needed a few disasters to get our teeth into – we needed a story, but right now, more than anything – we needed a boat!

So there we were in this white plastic bath tub, plus all the rigging, racing like river police to the starting line, a mile away, to qualify. We moved into the teeth of the armada amidst jeers which fell from our suntan-oiled skins like beads of sweat on a fat Hawaiian Wahine's big, brown back.

The competition official was standing on the pier, holding his official papers as he ticked off the starters on the line. Alex was standing next to him. Aha – our sixth crew member – was there, but I was still the unknown quantity. There had been talk of disqualification due to crew changes, but no one had been sure – not until now that is.

The crew change was declared and reasons given. All good, sound ones. Mr Mitchell was called away on business. The world must go on and Mr Steadman had kindly consented to take his place. Flew all the way from England in fact.

The refusal was flat and ominously reinforced with talk of rules and regulations as though King Kamehameha the Great had, himself, made them up and curses would fall on those who broke them. This was Hawaii after all and everything here was governed by dark and mysterious forces.

Then, someone mentioned an entrance fee of $100; it may have been the official. 'That's the only way,' he said, even though the full fee had already been paid – plus, of course, $4,000 to the 'ace' mechanic who couldn't get the *Haere Maru* going on time.

Nonetheless, all eyes were on me and, like one who knew his fate was sealed, as well as for the sake of the 'story', I fingered my wallet and looked at Hunter – 'I suppose it's all part of the story, right, Hunter?'

'Ah! The story – it's a pity you mentioned that. Yes, the story. Well, okay, but it's up to you.'

I had enough belief in the failure of this story to know it was pointless to back out now. I handed over the cash. Not so much as a receipt came my way. That official pocketed that bill there and then, as surely as the island was a living volcano.

So we were off.

'Welcome aboard again, Ralph,' said Alex, mincing around as though Queen Victoria was still on the throne of England and this was Henley Regatta. As I emphasize, I'm not a good sailor and hardly three hundred yards out, my stomach was beginning to moan – Oh! Christ!!!

This was an entirely different day. Yesterday was a picnic, and a signal to Captain Steve that something had not been quite right with his engine. But the swine wasn't going to give us the bottom line on why he had spent half the night trying to fix the problem.

The other crew members immediately began to prepare the tackle for trawling. The long poles called outriggers are turned outwards away from the hull of the ship and four rods thrust backwards from fixing holes along the stern. The out-riggers act as flag poles and hold a line high in the air, clear of the ship's swell. Lures in fancy rainbow colours are attached to the lines of the four rods and thrown in to spin wildly through the water, the idea being that big fish beneath us, bored with the lack of action down below would see these lures and swoop on them like eagles, gobbling the lot, hooks and all.

That's where I came in. As soon as there was a strike and our boat was 'hooked up', as fishermen put it, I, in my capacity as 'chair-man', had to leap for my harness – a blue waistcoat with straps and clasps which snap onto the rod which has the fish on it. The base of the rod fitted into a bell-topped tube below the fighting chair and I was in business. I was to fight the fish till the poor bugger was so exhausted I had him at my mercy and that's when the gaffer comes in. Hunter, the gaffer in this crew – though two gaffers are often needed for a big one – was supposed to lean over the side and – wap! – drive the nasty barbed spike in, somewhere behind the gills of the stricken beast.

Missing it could often result in another desperate fight for the chair-man, who would then have to overcome the fish's will to survive yet again and coax it back to the side of the boat. It could be a messy business, so one didn't want to be saddled with a sloppy gaffer. It could also seriously overtax the chair-man's art. Art and fatigue are not happy bedfellows.

So we trawled. Four hours went by and still we trawled. I fell asleep and woke only to find half the crew trying to crush killer bees which had somehow found our boat attractive – like seagulls following a liner for scraps.

A picnic was prepared with great expertise by Hunter and Alex. Part of the ritual of this was stringing it out on the assumption that just as the meat, mayonnaise, pickle, onion, and so on, had been laid carefully on rounds of bread, there would be a strike and the whole lot would fly in all directions. But who cared if you got the strike? That's what a fisherman is actually there for.

However, there was no strike and the sandwiches went

down well with beer. I had overcome my sickness by this time. Maybe the sleep had done the trick.

The afternoon wore on and we got further and further south. Lazy jokes and daft remarks had taken the place of conversation. It was obvious to anyone that nothing was going to happen, but a fisherman's most prized quality is his optimism. Many times that afternoon I sat in the chair and pierced the water's surface with my artist's eyes and my urging will.

Captain Steve was quiet as he turned the boat around in a wide arc.

'I guess we should've gone north,' was all I remember him saying during all those hours.

Hunter had by now begun to work. A notion that 'the worst fishing stories ever told' could be a good theme had started his brain off. The one we were in was so obviously the worst one yet . . . I looked over his shoulder to read some of his notes. After all, I was going to illustrate them? I read . . .

Notes on Notes about Notes about Tapes:
This is no joke.
It sounds crazy.
It sounds impossible.

White cockroaches.
Cockroaches survive Nuclear war . . .

And stuff . . . stuff! Hmmm.

Everybody has the same chance but you need to be
rich . . .

. . . Snap off the leaves of every . . . tree on the island
as a protection.
Why are there no Negro Fishermen? . . .
. . . Only a madman blows the whistle . . . (Everyone
on board knows the truth, but nobody will say it . . .
Maybe it takes abject despair to . . . realize
Something must be done . . .
Live bait – Sadie's dog. Customs won't let you take the
dog back to England . . .
. . . Thoughts come and go like boat swell . . . and the
weird man sleeps . . .
. . . the sea shall not have them . . .
Only a madman goes to sea in a beautiful pea-green
boat . . .
. . . Lies are bait, if you believe . . .
. . . It's on the line! . . .
. . . Start fishing, two hundred!! . . .
. . . Nothing . . .
. . . How I survived 100 days on frozen lettuce in an
open boat . . .
. . . The shame of the human race is . . .
. . . A floating death . . . and a . . . living humiliation
. . .
. . . There are no fish – do you speak English? . . .
there are no fish . . .
. . . Rich boys always win . . .
. . . The chartered boat and the sea . . .

'Wake UP!'
'What?'

'You were sleeping!'

'I was fishing.'

'Only in your dreams, Ralph.'

'I had one.'

'You were had.'

'It wasn't my money.'

'You were lucky.'

'I only wanted a fish.'

'You're asking for the moon!'

'Any minute now, I'll pick it out of the water and give it to you, with the Captain's compliments, Hunter.'

'You're drunk, Ralph . . .'

The afternoon was getting late and we were now stooging north. I didn't realize how far south we had gone. We were miles from Kona. Clouds had formed, the sea was choppy and wave slopes were slapping our boat with a little more intensity now as we headed directly into a southerly headwind. We were still trawling lures behind us but we had all but forgotten them. Talk had become sloppier and derogatory. Six men in a boat and no fish all day was beginning to tell.

The radio occasionally crackled results of catches. Someone either knew what they were doing, had struck lucky or simply went north. Hunter was drinking Wild Turkey and I joined him to lighten the load of boredom. Maybe it was then we got the idea to raise a black flag and steam into harbour with it flying. I found a black T-shirt and tied it to the outrigger. Our only offering was a full toilet can. Why not drop it in the harbour? That was how low we had sunk by that time.

It was all a bitter disappointment to Captain Steve, but he still talked of 'maybe' and 'you guys have gotta come back again. I'm going to catch you a big one. Slur . . .' It was too late – no more bullshit. This had done it. Even the lures were pulled in – it was over. Then it wasn't and oh, hell, no – the engine failed – started again and died. Out of gas.

It was the last straw. The boat rocked uneasily and dangerously, completely without direction. The final insult. A boat does funny things without its propulsion and a crew immediately senses its helplessness a mile out from shore.

Steve was trying to contact someone in the area and, as luck would have it, and that's what we were all praying for right now, an old buddy called Spalding was less than half a mile way and coming in our direction. Alex, the second chairman, had only three hours before thrown a Heineken bottle in the direction of their boat — a joke, admittedly, but not one to have been taken the wrong way.

The boat steamed up and its crew, four strapping, fat, brown and diehard seadogs, called for a rope to be thrown over. Captain Steve had no rope. No rope! On a small bucket of a boat that could sink at any moment if the engine was not working and he had no rope!

Thank God for Spalding. Most — in fact, all — of us, knew what his private thoughts were at that point — and I doubt if any of us had supreme confidence in our fate. Chipper who had been silent for most of the trip, leapt to the front of the boat to catch a line proffered by the crew of the *Keiki Iki* (meaning 'small child'), Spalding's boat. He lashed it to the front tie-lugs and we were pulled very slowly, but, nevertheless perceptibly, forward. The relief in all of us was not registered externally but you could sense it.

The *Small Child* took the strain and chugged homewards. We relaxed a little and went quiet. Then the rope snapped. Prayers were said. Uncertain laughter, uneasy forced laughter, hollow lame jokes and sick thoughts filled our minds, numb though we were.

Suddenly a fresh line appeared on the *Small Child* as though from nowhere, like the sun coming up, a bright sharp roll of rope. The sight of it was wonderful.

The same procedure followed. Breaths were held — would it hold? The waves beat on our hull more strongly than before — the swell was up as the early evening came on. Maybe later the sea would calm but right now later seemed an age away and possibly *too* late.

Fervent prayers poured through my mind and through the minds of the crew too, I reckoned, judging by the sickening banter. We got closer and closer to the port. Why not just go in? I kept asking — why go along the coast? What's wrong with dry land? But no! We must get back to port, the harbour where we had begun our ordeal.

There was no shame in our Captain. Couldn't he just cut his losses and make for shore – save his crew – and lose his own face? However, the complexity of his seafaring mind was beyond understanding and he was not bending to wind or circumstance.

As we neared the protruding point before the harbour, we had to slow down to make a turn into the narrow entrance. Our rope was obviously too long and the danger of whiplash caused by current or sudden wave movement was clear, even to a landlubber like me. Hunter shouted as much and threw a curse towards the *Small Child* to give it greater emphasis.

'Shorten the rope! Shorten the rope! Goddamn you!' he screamed.

The crew were gripped in a state of shock which was apparent from their infantile giggling. Something must have registered with Spalding for he immediately began to pull us in – pull us, that is, close, berth to berth – grinding and jamming with the wave movement which was as treacherous and unpredictable as waves on a jagged shore line.

Sense, for some unknown reason, prevailed – or such was the bungling that pure chance found us away from each other and pulled closer but safely through the mouth of the harbour. The waters calmed and we were safe inside the harbour wall.

We were towed towards the fill-up point where we should have filled up properly in the first place. Spalding cut us loose and we careened towards the jetty, dangerously close to crushing someone's fibreglass dinghy. Its owners had anticipated the possibility and were frantically trying to push us away,

'Mind our boat – mind our boat!'

Sundowners were scattered as we ground into a tight area – bums in the air trying to fend off impending disaster. Fate was kind yet again and, miraculously, nothing was crushed and nobody fell overboard. However, my camera did. Yes! My camera!! My beloved Pentax SP 500 with the Super Tacumar 1:2/55mm went over and right to the bottom. A local kid offered to get it for me for ten dollars and I was so happy. The camera was retrieved but it never worked again. I still have its corroding body in my studio. I had the film processed, the images like wrecks in muddy water. Hunter and I picked up our baggage and leaped like antelopes for the jetty, perhaps to hide our shame. Our car was parked up on the rise.

That was it, no more sea, ever. At least for me. But Hunter had other ideas.

Tuesday, 26 May
Nostalgic notes from a cold climate
[I wrote these notes back at home after our disappointment, perhaps trying to find rhyme or reason for pursuing the idea of another attempt.]

And yes − I'm beginning to love Hawaii. I never could, of course, but yes, it's coming upon me like a beauty queen looking for a sponsor. The place is sweaty and against my natural inclinations. Moments after arriving, one feels trapped but irrevocably part of its history. It goes against the grain of all I ever want in my work. Against my European roots. Yet I need to go back. I need the burns. The helpless, awesome emptiness of the place gives me a chance. There is something there waiting to be solved. There is such a lot of cement holding me in Europe. I could not work in Hawaii. I would go into real estate. I would lose myself in the soft, new way.

There isn't a soft, new way. It's hard and mean but there is a chance, and that is all one needs. Local inhabitants who I know in my heart are less than me, goad me into becoming a part of them.

I can teach them something they need to know but they could teach me so much more. I have a feeling that there is no way back or that may be a good feeling. The islands are unique and overrated but within their vulnerability lies a deep call.

I feel myself wanting them to demand my return. I think it is a mistake but there is really nothing here

I can relate to any more. I feel lost and not a part of my English way of life.

Hawaii is not trying to be the centre of the world. It has a cheapness I despise and a worthwhile cheapness. Cheapness is what we have all been striving for. Cheap like we are. Cheap style and cheap tricks. It is not worthwhile but it feels like life. Quality is for the gods and I am not a god.

Oh, God have mercy on all us poor wretches. We are only your children and we know what we do.

Before we left we made a visit . . .

Return to the City of Refuge to seek and find a new religion

The last day of my short, second visit Hunter and I had to talk. The only place to go now was the City of Refuge. Away from this sham.

Somewhere we could mull over our experiences, away from the feckless ways of man, particularly Kona man. Whether we went there to lick our wounds or to hatch a plot is uncertain. We were mortals among gods and it hardly showed. Not then.

The City of Refuge reverberates with wisdom of time and ages past. The legend goes that if a man can hide out inside the City of Refuge for seven days, he can enter the outside world again, an innocent man, no matter what he has done. I had hardly seven hours before taking the flight back home. One's every footstep seems recorded and marked for future reference.

The sacredness oozes out of every black crevice. The black rock is sacred and jealously guarded by the Goddess Pele, Goddess of the Volcanoes, crossed in love and as schizophrenic as three madmen. Black rock must never be taken from the island under pain or misfortune. Where was our story now? It hardly mattered. The most important thing about being in the City of Refuge is simply that you are there, steeped in the blackness and fanned by the waving palms, surrounded by stories, far more significant than ours. They are stories that speak with authority, tried and tested and passed down from father to son and mother to daughter.

We had no story to tell there. There are no stories to tell gods. They know them all. We need only listen and look. Their stories are everywhere. Carved grimacing faces shout truths that never change and stand as irrefutable proof of what no longer survives outside this sacred place.

Weathered, carved figures towered above us, rooted to this hallowed spot, a shrine of simple beliefs and dark legend protected by nothing more than a coarse, thatched roof and a fence of spiked, wooden posts forming a barrier around the shrine and a warning to mortal man to stay outside.

There is a small door at floor level on one side, about two-and-a-half feet high, guarded on its right by two fearsome totems. There was a fascination for that door which was difficult to resist, like the compulsion one has to jump when standing on a cliff edge. What in god's name was inside? But we did resist. We both feared the worst, but our curiosity to know the unknown was intense. What on earth could be inside a hut so rude that it needs to be protected by such awesome effigies which are both man and beast and yet sacred god figures to those who believe?

There is a passage of explanation in Freud's *Totem and Taboo* which says that taboo remains a power because it is a power from a mental conservatism and forms the basis of our moral attitudes and laws and that must account for the reluctance to enter in. To overcome our fear of mere superstition is a super-human effort because of the 'if' factor.

In tribal life, Freud goes on, the father figure is all-powerful and all the women are his. The sons of the tribe grow restless and jealous. They rise up and kill the father so that they can take the women for themselves. Such is their remorse that to appease their guilt they re-create the father in the form of a totem or god, then bow down and worship him.

So father figure is a god. A hero who must suffer in order to appease the 'tragic guilt' of the horde, achieved by a process of systematic distortion or refined hypocrisy. This, of course, is ultimately religion.

There was nothing inside that hut, but neither of us would enter because neither of us could overcome our fear of 'if.' Instead, I played the cheapest trick of all. Modern man's cheapest trick, that is. I nipped over the spiked, wooden fence and Hunter, now accomplice to this act, took a Polaroid picture of me standing with the gods — our fathers. Then I nipped back quickly because the longer I stayed on such supernaturally forbidden territory, the shorter the odds would become.

To primitive man, wishes and impulses have the full value of facts and in our own minds are governed in some measure by that. The uninhibited thought or the unthinkable passed directly into action, albeit half the action, but nevertheless we both experienced fear. There is nothing inside that hut but our darkest fears and, to the Hawaiian, his guilt since time immemorial. It is a monumental lie systematically told and retold and distorted — refined hypocrisy. A legend with the full force of fact and the blessing of time.

Hunter and I left the City of Refuge better men than when we arrived. We went to ask the question: 'Why do they lie to us?' We goaded the gods — just a little — not too much. And we both knew the answer, each in his own way. The gods told us. They laughed at us and mocked our puny efforts. They laughed at our lies because they were so much greater.

When I boarded Aloha Airlines, Flight No. AQ88, I turned for a moment, long enough to take a small piece of black Kona rock from my pocket and hurl it onto the tarmac where it belonged. That would have

been goading the gods just a little too far, even though
it is probably only a lie perpetuated by the hordes of
real estate agents who guard their weird investments
with all the power of gods at their command.

And they have the gods' blessing. It is all part of
the legend — and the lie.

Hunter stayed on fishing and wrote me letters about his
progress, then sent me pictures of a triumphant catch with a
caption which said: 'Ralph! We killed like champions.' I
replied:

    9.6.1981
    Tentative notes from a loser to a cham-
    pion.
    Dear Hunter & Laila
    First and foremost, congratulations on your
    stupendous catch!!! I find it hard to believe.
    Harder even because all these things happened
    after I have left.
    I left for good, solid, family reasons,
    but I could have stayed if you had guaran-
    teed the fish. It really is a tough life.
    Perhaps that is why I am such a limp person.
    Everything happens later. It happened in
    Washington the same. The big fish came after
    I left.
    But leaving is an art form too, for good-
    ness sake. Leaving at the precise moment for
    full dramatic effect. Pure theatre. An exit
    is often more impressive than an entrance.
    Famous last words prove that. I never heard
    of any famous birth quotes. Everyone is the
    same at birth - just a squealing mass of
    survival. But after a life of experience and
    pain one knows what to say.
    I enclose my diary of events from my work
    in Kona. It has its moments and is only meant

as a memory-jogger for you. What you may have
missed - or wish to be reminded of - or even
forget. I particularly like the Cry of Refuge
piece and feel it is unfinished but the point
is made. There will be drawings about that.
It is for me, the most fascinating aspect of
our visit at any time.

But the fishing story we have spoken about
and the things which surround it are, without
doubt, a marvellous yarn - but they need the
storyteller's touch - and it's yours! As I
said on the phone, that seems to be what the
world needs now. Just a good story well told
- like the kind that used to be told around
firesides on winter evenings in places like
Copenhagen and Koblenz - two places we still
have to go to for marathons. (Whatever happened
to Running after I refused to do the London
Marathon - where are they?)

Are we merely the barrels sawn off and
discarded from a 12 bore shotgun?? I keep
asking myself. Where are they now? Those people
who believe we have something to say. Do we
merely speak for a generation or do we speak
for mankind? Are there friends out there?

Well, there's Alan (Rinzler - first editor
to attempt to make Lono a reality) and I wish
you every urging persuasion on June 15th, when
he arrives to work with you. My end will be
backed to the hilt. You have no fear of that.
I intend to give him more than he ever hoped
for. Always give more than is expected. It's
the only ace in the pack. And do some bloody
drawings while you're at it, too! You saw the
actual fish. What did it look like in the
water? Did you see its eye? Did it look at
you? You personally? Most victims are supposed
to look their victor in the eye before they

die and ask a profound and silent question
which leaves a deep impression on the one left
behind.

Anna has just confounded me with a
distressful revelation — that I did not say
hello to her mother's friend in Canterbury
today. She is, so Anna says, in tears.

I am a bastard. That's the cold uncommon
truth and I do not know how to repair, or I
am too far gone to bother.

Over to you. This world is a real puzzler.
God bless
Ralph

We wrote to each other many more times, and often at
length, during the tortured creation of *The Curse of Lono.*

But there is one thing you should know,
Ralph, before you take your theory any further:
I am Lono.

Yeah. That's me, Ralph. I am the one they've
been waiting for all these years. Captain Cook

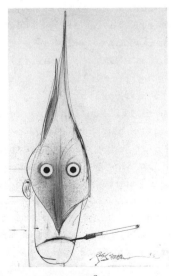

was just another drunken sailor who got lucky in the South Seas.

Or maybe not - and this gets into religion and the realm of the mystic, so I want you to listen carefully; because you alone might understand the full and terrible meaning of it.

A quick look back to the origins of this saga will raise, I'm sure, the same inescapable questions in your mind that it did in mine, for a while . . .

Think back on it, Ralph - how did this thing happen? What mix of queer and (until now) hopelessly confused reasons brought me to Kona in the first place? What awful power was it that caused me - after years of refusing all (& even the most lucrative) magazine assignments as cheap and unworthy - to suddenly agree to cover the Honolulu Marathon for one of the most obscure magazines in the history of publishing? I could have gone off with a plane-load of reporters to roam the world with Alexander Haig, or down to Plains for a talk with Jimmy Carter. There were many things to write, for many people and many dollars - but I spurned them all, until the strange call came from Hawaii.

And then I persuaded you, Ralph - my smartest friend - to not only come with me, but to bring your whole family halfway around the world from London, for no good or rational reason, to spend what might turn out to be the weirdest month of our lives on a treacherous pile of black lava rocks called the Kona Coast . . . and then to come back again six months later, at your own expense, for something as dumb and silly as the Jackpot Fishing Tournament.

Strange, eh?

But not really. Not when I look back on it all & finally see the pattern . . . which was not so clearly apparent to me then, as it is now, and that's why I never mentioned these things to you in Kona. We had enough problems, as I recall.

[. . .]

I am Lono. And you're not . . . Indeed, and that explains a lot of things, eh?

[. . .]

They can't touch me now, Ralph. I am in here with a battery-powered typewriter, two blankets from the Bali Kai, my miner's headlamp, a kitbag full of speed and other vitals, and my fine Samoan war club. Laila brings me food and whiskey twice a day, and the natives send me women. But they won't come into the hut — for the same reason nobody else will — so I have to sneak out at night and fuck them out there on the black rocks. We scream a lot, but . . .

So what? It's not a bad life. I like it in here. But I can't leave, because they're waiting for me out there by the parking lot. The natives won't let them come any closer. They killed me once, and they're not about to do it again.

Because I am Lono, and as long as I stay in The City those lying swine can't touch me.

It's a queer life, for sure, but right now it's all I have. Last night, around midnight, I heard somebody scratching on the thatch and then a female voice whispered, 'When the going gets weird, the Weird turn pro.'

'That's right!' I shouted. 'I love you!'

There was no reply. Only the sound of this

vast and bottomless sea, which talks to me
every night, and makes me smile in my sleep.
  Hunter
  (Kick ass!)

I replied:

  9th July 1981
  Chants and incantations from the Altar of
Kent and strange rumblings from the cesspit
down the valley. (re. The Curse of Lono).
  Hunter,
  Just when I had finally laid to rest the
last remnants of belief in any god from my
soul, your letter arrived and poured scorn on
my faithless heart, once and for all.
  If you are indeed the god Lono, and all
the signs point in that direction, then I am
your high priest. I am the purveyor of tidings
regarding your whims and your wishes. I am he
who runs barefoot through the scrub on the
big island, proclaiming your message not in
words - for these natives, the true ones, the
descendants, cannot read and must be shown in
pictures. That is my true metier, my craft,
and the one which will finally and clearly
convey the message of your coming in a manner
that no fear riddled real estate merchant can
any longer deny.
  These people, your people, the true
Hawaiians, are hungry for their past, their
true destiny and you have inspired the hope
in their souls. They want to rid themselves
of the yoke of blatant capitalism once and
for all and re-establish their noble ties with
their homeland. And who can blame them?
  I have known about this for longer than
you realize and the vague germ of the idea

has finally laid an egg. I enclose two black
and white photos of you in characteristic pose
and when I first saw them myself the 'god'
thought struck me then and from them you will
see why. I don't know what brilliant flash of
inspiration caused Laila to buy you that Samoan
War Club - she's an extremely bright and
perceptive girl (she can even tell from 20
yards that I was wearing underpants beneath
my blue shorts!), but it is now so much of
the Lono story, let's even consider it as
cover material.

As for more mundane matters, an art director
is the last person we need to bring in just
now. I will handle the majority of that with
you and if she is willing - Laila?

At that point I should come to Aspen - or
you to Kent with Laila. Whichever way we can
only finish it together where we can commu-
nicate easily, add drawings, change them, add
text, change it, place it in a certain context
and whatever. It requires a close liaison and
cannot be done from 6000 miles. You have
cracked a real bastard here, mate and we're
all proud of you. Sadie too! She knows the
fish is not plastic. I've explained it to her.
It's not the one you hire from a glass case
in the Kona Inn for such occasions.

Much of the layout incidentally will require
my distinctive handwriting - and yours. Chapter
headings, titling, etc. It is a necessary part
of the characteristic style of the subject.
     Best. Ralph

While writing this book, I have set about reading everything
else we have ever written to each other. Hunter could be solici-
tous, caring and, above all, funny. When, for instance, one of
my sons got into trouble in late 1981:

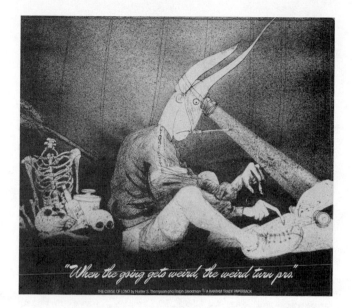

"When the going gets weird, the weird turn pro."

THE CURSE OF LONO by Hunter S. Thompson and Ralph Steadman © A BANTAM TRADE PAPERBACK

17.11.81

Lamentations of a failed father — Glue sniffing in a fractured family — time passes as brains go soft and love for the accused is strained to fever pitch.

My son has been picked up by the police with another brick in his hand. The other one was already through a $500 plate glass window. He also finds your book 'Hells Angels' fascinating, but doesn't care much for 'Fear & Loathing in Las Vegas' because you don't dwell at length on petroleum-based substances. He's nearly 16 and locks the bathroom door. When he comes out he leaves the wall heater on & opens the bathroom window. He denies this flatly even though I hang about outside the bathroom door until he comes out.

I beat him to within an inch of un-consciousness and still he denies it.

As a concerned Protestant father what should I do next? How can I stand by and watch him destroy himself - and more important, the family name???

I confiscated his gun because he shoots at children. He demands the gun with menaces but would settle for the money.

Juan was never like this, was he?

Are we the first generation of parents to spawn a mutant tribe? Have we taught them tricks even we would rather forget?

Is it because they don't believe in anything anymore & is the white man fucked?

Are the sins of the fathers visited immediately on the sons or aren't they supposed to wait a generation or two?

And finally - why me????

Don't try to answer any of these questions because you can't. You, like me, have nothing to say, no right to explain and certainly no ability to understand.

I rue the day I gave my son the justification to call me Judas.

So there it is - and we await the outcome. At least it's not theft or rape.

Yeh! God bless, send word or wire.

Ralph.

I received this wonderful, funny reply:

Dear Ralph,

I received yr. tragic letter about yr. savage, glue-sniffing son & read it while eating breakfast at 4.30 A.M. in a Waffle House on the edge of Mobile Bay . . . and I made some notes on yr. problem, at the time, but they are not the notes that any decent man would want to send to a friend . . . So I put them away until I could bring a bit more concentration to bear on the matter . . .

And I have come to this conclusion:

Send the crazy little bugger to Australia. We can get him a job herding sheep somewhere deep in the outback, and that will straighten him out for sure; or at least it will keep him busy.

England is the wrong place for a boy who wants to smash windows. Because he's right, of course. He should smash windows. Anybody growing up in England today without a serious urge to smash windows is probably too dumb to help.

You are reaping the whirlwind, Ralph. Where in the name of art or anything else did you ever see anything that said you could draw queer pictures of the Prime Minister and call her worse than a de-natured pig, but yr. own son shouldn't want to smash windows?

We are not privy to that level of logic, Ralph. They don't even teach it at Oxford.

My own son, thank god, is a calm & rational boy who is even now filling out his applications to Yale & various other eastern, elitist schools . . . and all he's cost me so far is a hellish drain of something like $10,000 a year just to keep him off the streets & away from the goddamn windows . . .

What do windows cost, Ralph? They were about $55 apiece when I used to smash them — even the big plate-glass-kind — but now they probably cost about $300 apiece. Which is cheap, when you think on it. A wild boy with a good arm could smash about thirty big plate-glass windows a year & still cost you less than $10,000 per annum.

Is that right? Are my figures correct?

Yeah. They are. If Juan smashed 30 big windows a year, I would still save $1000.

So send me the boy, Ralph - along with a
certified cheque for $10,000 - and I'll turn
him into a walking profit-machine. Indeed.
Send me all of those angry little limey bastards
you can round up. We can do business on this
score. Just ship them over, with a $10K cheque
for each one, and after that you can go about
yr. filthy destructive business with a clear
conscience.

The Prime Minister is a de-natured pig,
Ralph, and you should beat her like a gong.
Draw horrible cartoons of the bitch, and sell
them for many dollars to The Times & Private
Eye . . . But don't come weeping to me when
your son takes it into his head to smash a
few windows. You might as well try to teach
a young dog not to piss on a tree.

Have you ever put a brick through a big
plate-glass window, Ralph? It makes a wonderful
goddamn noise, and the people inside run around
like rats in a firestorm. It's fun, Ralph,
and a bargain at any price.

What the fuck do you think we've been doing
all these years? Do you think you were getting
paid for yr. goddamn silly art?

No, Ralph. You were getting paid to smash
windows. And that is an art in itself. The
trick is getting paid for it.

What? Hello? Are you still there, Ralph?

You snivelling, hypocritical bastard. If
yr. son had your instincts, he'd be shooting
at the Prime Minister, instead of just smashing
windows.

Are you ready for that? How are you going
to feel when you wake up one of these mornings & flip on the Telly at the Old Manor
just in time to catch a news bulletin about
the Prime Minister being shot through the

gizzard in Piccadilly Square . . . and then some BBC hotrod comes up with exclusive pictures of the dirty freak who did it, and he turns out to be your son?

Think about it, Ralph; and don't bother me anymore with yr. minor problems — Just send the boy over to me, I'll soften him up with trench-work until his green card runs out, then we'll move him to Australia. And five years from now, you'll get an invitation to a wedding at a sheep-ranch in Perth . . .

And so much for that, Ralph. We have our own problems to deal with. Children are like TV sets. When they start acting weird, whack them across the eyes with a big rubber basket-ball shoe.

How's that for wisdom?

Something wrong with it?

No. I don't think so. Today's plate-glass window is tomorrow's BBC story. Keep that in mind & you won't go wrong. Just send me the boys and the cheques . . .

(I can't spell that word, Ralph; but I think you know what I mean. It's what happens when the son of a famous English artist shows up on the Telly with a burp-gun in his hand & the still-twitching body of the Prime Minister at his feet . . .)

You can't even run from that one, Ralph — much less hide, so if you think it's a real possibility all I can advise you to do is stock up on whiskey and codeine. That will keep you dumb enough to handle the shock when that ratchet-head, glue crazy little freak finally does the deed . . .

The subsequent publicity will be a night-mare. But don't worry — Your friends will stand behind you. I'll catch one of those

Polar Flights out of Denver & be there eight
hours after it happens. We'll have a monster
press conference in the lobby of Brown's Hotel.

Say nothing until I get there. Don't even
claim bloodlines with the boy. Say nothing.
I'll talk to the press. And we will bury your
shame forever, in a blizzard of angry
bullshit.

Right. And how's that for art?

Never mind. Let's get back to this terrible
problem you're having with your son. He's a
murderous little bastard, for sure . . . and
. . . Jesus, Ralph, I think I might have
misspoke myself when I said ten thousand would
cover it.

No. Let's talk about thirty, Ralph. You've
got a real monster on your hands. I wouldn't
touch him for less than thirty.

[Handwritten: Whoops — I just got a call
with regard to the opening of 'F&L in Las
Vegas' in London on Jan 25 — where I will be
the guest of honor.]

You're in luck, Ralph. I can counsel the
boy personally in my suite at Brown's Hotel.

I can film my personal counselling sessions,
as well as the stage production.

See you soon,

Yr. buddy HST

TAXI TO THE SUBURBS
1981

No room at the inn ... A savage encounter on a stairway in the dark ... Terror hitches a lift on the Battersea Park Road

In 1981, a dramatized version of *Fear and Loathing in Las Vegas* was staged in London. I was commissioned by *Time Out* to write a piece about it. Hunter was unhappy about it.

```
Ralph,
    Before we go into town today you should
get ready to explain to Deborah Rogers just
exactly how you secured British Rights to Vegas
for Lou Stein.
    Nobody secured British Rights for anything
- & this question will be resolved today.
    I want $1000 for every day the play runs
without permission - and without any signed
agreement with me - the Author -
    Uncle Ralph has made his deal - now let
him collect on it.
    Stein should have known better.
    Call P. Morrisey's lawyer on this
    I will not go near the Gate Theatre until
this issue is resolved
    See Bob Musel AP - & explain to Ateyl &
get out of town
    ASAP. H
```

I must've been early getting to the Latchmere Theatre that Monday. The theatre was situated above a recently renovated pub along Battersea's conveyor-belt main street. I was wearing my Allen Ginsberg skull-mask with glasses to avoid unwanted confrontations. I found the back entrance up a newly erected iron fire escape and went inside.

Immediately on my left, there was a door. It was open. It looked like a toilet. Inside, a dark figure hunched over the pan. My god, I thought, I hadn't expected such realism, not so soon anyway. I've been out in the country for two years, so I'm out of touch with theatre. I didn't like to ask if he was okay. Actors get funny at rehearsals, though I didn't remember a bag of plumber's tools lying there in the actual book. So I moved on.

Stumbling down a dark corridor, over pipes and dangerously loose wires I made my way toward sounds of banging and out-of-tune whistles.

'Ah hullo!' I grabbed the hand of a tall, slender character emerging from a door with light beyond. He was wearing shades. I shook his hand warmly.

'My God, you look just like him,' I said. He looked at me and he looked at the mask I was wearing. 'Oh, sorry, don't worry,' I said as I removed it. 'It's me, Ralpho Steadman. You must be playing Hunter S. Thompson. It's amazing! You could be him.'

'What?' He looked at me strangely.

'Hunter.' I continued, 'You've got him off to a tee, the way you move, the nervous twitch. The weird, deep voice . . . Amazing.'

'Er . . .' He looked a little nonplussed.

'Oh sorry,' I said. 'Relax, it's okay. I'm not spying. Just looking in. You know. You're great — keep it up. Ah . . . don't I know your face. From telly perhaps? Great stuff.'

'Don't think so,' he said shuffling uneasily.

'Oh sorry. It's theatre, right?' I replied looking for openers. 'Real acting! None of those crappy retakes. Straight off first time. Know your lines. Tradition. That sort of thing. What's your name? I ought to know you.'

'It's Arnold,' he said. 'I'm the electrician.'

'Ah . . . yes — just the man I want to see. Look — er, my wife's thinking of opening a nursery school in our outbuilding.

Nothing grand. About thirty kids. Proper thing. Not just child-minding. The place needs electrifying of course. You ever work out of town? Never mind. I'm looking for Lou Stein. Is he here?'

'Try up those stairs. Take a left, along the corridor, then right down the end. They'll know.'

'Thanks. And don't forget — day in the country — do you good.' I scribbled out my number and turned to leap upstairs.

'Oopsaaaieeeeeeee!' A sharp pain, like thirteen cattle prods all at once, took my breath away as I drove my shin bone into the bonnet of a huge, red fibreglass Chevy convertible being carried across my path.

'Ouch and hell!' I screamed, as I tried to hold in the agony and sorrows I'd long forgotten. 'Sorry, my fault! Is that it? The red convertible? Oh no! Arrrgh! No!'

'Are you all right?' The voice came from a tiny girl in a cream-white fluffy V-neck sweater who had been carrying the Chevy with another figure in the shadows.

'Blood!' I screamed. 'I'm bleeding! Oh God, all over my pants. Get me a doctor quick! A real doctor. Nothing Gonzo — I don't want to die . . . !'

'Relax, man. It's only red paint. We just sprayed the Chevy.'

'Thank god!' I said, then stiffened. 'My pants!' They're new pants! What about my pants? I wore them specially. First time on. Okay, okay, never mind. It wasn't your fault.' Calming down, I said between gritted teeth, 'I'm looking for Lou Stein. Thanks,' I said as she pointed upstairs. I put my mask on again and made for the stairway.

'Mind your hea —!' said the girl in the cream-white V-neck. 'Oh dear, people are always doing that.'

Dazed but still in control, I stumbled forward along the gloomy landing above towards a crack of light.

I knocked and pushed open the door. Two girls sat on boxes at a desk that these days you can only find on out-of-town rubbish tips; the kind that fetch big prices in furniture caves.

'Hello!' I said. 'Is Lou here? He's expecting me.'

'He's up at the Gate, rehearsing.'

'He said he'd be rehearsing here.'

'Well he's not due here until tomorrow or Wednesday.'

'Shit! I came here specially. I'm reviewing the play for *Time Out*. My career depends upon it.'

The girl nearest the window looked out at the rush hour traffic down below and picked up a magazine.

I could tell their coffee was cold by the way the milk was floating on the surface. They couldn't possibly know how many sacrifices I had made to be here on time. That feeling rises in me more often than it used to when I come to London. It's probably just me.

'They'll be rehearsing up there if you want to go up now. We can't tell him you're coming. We're having trouble with the phones.'

'Can't pay the bill eh? Heh! Heh!' My tongue crawled around the words behind my Allen Ginsberg mask.

'Not at all. There's trouble on the line. You know what Telecom's like! You can pay but you can't say.' That's good, I thought. This kid could be smart one day. She can't be more than fifteen and a half. A helpless enthusiast.

'Can you get me a taxi? It's ten past five already.'

Naturally, I was down on the Battersea Park Road within minutes waiting for the cab. The rush hour was building up. Sarah, the young lady with me, a *Fear and Loathing* student no less, was by this time doubting my ability to handle even this simple operation.

'I have always found,' I explained, 'that if you have ordered something to happen like a cab downstairs in five minutes, then you should abide by that. Otherwise, where would we be? These people would never cross London for anybody. They are scum. So think of those who come after. It's the only decent thing to do!'

'Bullshit!' she replied. 'If those motherfuckers aren't here like they should be, why wait?'

The words left her petal-lips like a burst of fire from a flame thrower. She was right, of course. What kind of service was this?

We hailed the next cab and got to the Gate Theatre in little more than fifteen minutes. Even the young can teach you something, crude as they are.

This time it was cut and dried. Upstairs we were greeted with knowing looks of friendliness. We were home at last.

Through the scenery sat a girl. We walked around the actors mouthing their lines, like late-comers to an evening performance.

Attorney: I want tacos . . .
Duke: Five for a buck, that's like . . . five hamburgers for a buck.
Attorney: No . . . don't judge a taco by its price.
Duke: You think you might make a deal?
Attorney: I might. There's a hamburger for 29 cents. Tacos are 29 cents. It's just a cheap place, that's all.
Duke: Go bargain with them . . .

We settled down comfortably, yet decently quiet. I searched frantically for the characters of a play. The two I knew best, that is. I saw Oscar immediately but where was Hunter? Maybe that was him with the hat. I took out a notebook and started to draw. It looked interesting, a group of people seated together, waiting expectantly for the prompt to give them their lines. Gradually I focused on the characters and figured out which one was meant to be Hunter. Of course, when I saw him it was obvious. Small in stature but where was the nobility? I looked in vain . . .

Waitress: May I help you?
Attorney: . . . Yeah, you have tacos here? Are they Mexican tacos or just regular tacos! I mean, do you have chili in them and things like that?
Waitress: We have cheese and lettuce, and we have sauce, you know, put on them.
Attorney: I mean do you guarantee that they are authentic Mexican tacos?
Waitress: I don't know. Hey Lou, do we have authentic Mexican tacos?
Woman's voice from the kitchen: What?
Lou: We have tacos. I don't know how Mexican they are.

Attorney: Yeah, well, I just want to make sure I get what I'm paying for. 'Cause they're five for a dollar? I'll take five of them.

Duke: Taco burger, what's that?

I leapt up. 'That's not Hunter,' I screamed. 'He doesn't talk like this.' I grimaced, the memory stuck in my mind.

'Holy shit . . . he doesn't say "Holy shit". He says . . . "Hoooly Shit"! It's important. It comes from the gut. Sorry I didn't mean to interrupt. Do carry on.' I sat down and refitted my Ginsberg mask. I fucked up again. What a shame. I had thought I'd try to create a decent impression. The actors continued as though nothing had happened, mouthing their lines, doing their best, trying valiantly to find characters that were still ten thousand miles away.

'You're talking too fast,' I said. 'Remember the heat. They're in the desert, right.' Oh, the heat, they remember the heat. 'And the flies, swarms of flies. The place. The place is . . . a concrete block with a gaudy cheap sign outside. Inside there can only be slobs. Slow-moving scum. There's grease on the tables. There's nothing to hurry for – slow down. Talking is an effort. You'll give your order and the waiter'll look at you for five minutes before even her eyelids move. Slowly down, then up again. Then she'll move her head around like an alligator beckoning its mate . . . Hunter's polite. He's always polite. The sight of this place would put a look of bruised dismay on his face, but he'd be polite. Particularly when there's no alternative. This won't be the time he'll beat the table with his fist and throw his tacos at the wall like a lunatic. He only does that sort of thing in airport lounges and Holiday Inns . . .'

'. . . He . . . er, sorry . . .' The director was beckoning me to keep quiet. The actors had crossed their arms and were looking at their watches. I looked down and scribbled a few aimless lines in an attempt to cover my shame.

'Relax into your parts.' The director was talking now. 'The audience wants to enjoy this play,' he was saying. 'They're already convinced so you don't need to hurry.'

'Right,' I interrupted. 'Lou's right there! This book's

already gone down in American literary history like *The Grapes of Wrath* and . . . *Peyton Place.* You're dealin' with a classic here, a luxury these days, I can tell you.'

'Er, Stew . . . take it from . . .' – looking down at the script – '"The Taco had meat in it". Okay, quiet everybody.'

The actors started up again, and trying not to fidget I rummaged around in my bag for another pen.

'Damn! My letters! Psst! Lou! Psst! Is there a postbox near here? Forgot to post my letters. Must go today.' He nodded with his head in his hands. I stumbled up and over the back of the seat and looked for a door.

'Psst! Lou! Psst! Lou!' Lou turned his head around very slowly, his face the face of a man who looked as if he'd been imprisoned in the Bastille for two years and was used to pain.

'You'll have to go that way,' he said laconically, pointing through the actors to an opening at the back of the stage.

'Back in a mo'! You fellas carry on.'

It didn't take long to find the postbox and I was back in no time, starting across the stage just as Duke was saying, 'Don't worry. I'm insured against all damages for only two dollars a day.' He said this as he was walking offstage.

'The walk's wrong!' I screamed, leaping up again. He doesn't walk like that. He walks like this.' And my poor-fool mind allowed my body to attempt a pathetic imitation of the man they were trying to portray. At that moment I knew I shouldn't have come back because I just couldn't stop myself. I grabbed a bag and walked with it like he would walk, explaining all the while that *this* is what he does . . . 'He holds a cigarette this way, hand raised slightly cupped on a vertical forearm with his body off-balance, like this. He jerks his head back when he lights up — like this — as if to focus on the tip, and don't forget the Zippo lighter. He holds a steering wheel like this down between the middle fingers of his free hand like this, cupped on the palm so's he can swivel it through three revolutions without taking his hand off that point of the wheel. And when he bellows to God or no one in particular, it sounds like a rhinoceros with a tank on its foot — Ahharrrrrrrrrrrgh!!!! Eeeeeeeeeee!oooooooo!' I forced a sickly smile here as the actors backed off against the wall and I continued '. . . and he's big and he's clumsy, but he's gentle and it *is* him we're talking about. It'll be Raoul Duke, but it's Hunter S. You can't disguise it. He's got one leg shorter than the other and he walks like this and, dramatically speaking, it's most important. It gives the man his lumbering dignity. It gives him all he needs to melt the heart of the hardest stewardess on the cheapest airline anywhere in America and that's all you need to know.'

I slumped down in an emotional heap and tried to crawl under a big bag full of stage props. Nothing would bring me out, finally hoist by my own gibberish! They looked at me, but said nothing.

By this time I was laughing crazily. But it made no difference. I was just another fucked-up cleric with a bad heart. Shit, they'll love this down at the Brown Palace. I took another big hit of the amyl, and by the time I got to the bar my heart was full of joy. I felt like a monster reincarnation of Horatio Alger . . . a Man on the Move, and just sick enough to be totally confident.

THE FISH HAVE
GONE SOUTH
1981–6

He who can, can . . . He who can't, write about it . . .
When is a book not a book? . . . Is fishwrap art? . . .
A second chance.

I had forgotten how eaten-up inside Hunter had become as
he bitched and railed against my involvement in anything,
particularly the last but necessary remnants and details of the
book *The Curse of Lono*. He wanted it all and he wanted to
forget that I had been there since day one. The 'Fear' took
hold of him in August 1981 after the Honolulu Marathon and
our month's stay over Christmas on the black island of Kona
in our compound of chalets.

We were there in order to get the pictures done and to allow
Hunter to figure out why on earth people run in the first place.
In an article entitled 'The Charge of the Weird Brigade' he
tried to find the answer to the question. The article evolved into
a book – because, almost without trying, the powerful ambience of such a place brought to mind the myths and legends
that have grown up there. It was dark and moody, treacherous
and relentless, humid, merciless and bad, and if that wasn't
enough, beautiful and exotic, seedy, damaged and subject to
storms at a moment's notice. Writing or drawing anything about
Hawaii on damp paper requires confidence and peace of mind.

Towards the end of the two-year period of struggle, grief,
frustration and empty rhetoric, he rang me and asked me to
write to Ian Ballantine, the book's new editor and publisher at
Bantam Books, to explain as clearly as I could what it seemed

Hunter was unable to do. I became the messenger for a while. My drawings for the book were complete and I was happy to do anything to force an agreeable end to the whole ugly journey.

Months before, I had received a practically incoherent note from Hunter. It read:

```
Old Manor. Feb. 6th '82. Ode to a Bad Idea.
    Do 2. 5. 6 words by tomorrow and get the
fuck out of town, and never explain why because
it doesn't,,,?%? I am in a foreign land and
I am dumb and I am too old to learn to be
English, or anything else. I was born in a
place that I cant recreate here. And why should
I? Because I know I can work somewhere and
this is not it. This is the wrong place for
me to be. Repeat. The wrong place to be. I
HAVE THE FEAR. REPEAT. I HAVE THE FEAR. Those
of us who have been here know what the fear
is. The others do not. Selah. May the god of
the dark forces give me the grace to over-
come my evil heritage. But probably not. So
I'm out of town anyway.
```

Letters moved back and forth between us. It was the most active time in our entire friendship for correspondence between us as Hunter struggled like a doomed bottle-nose whale, stranded up the Thames, to nail this strange story of fishing and gods. I wrote and told him how upset Sadie was because he had thrown the pink radio he had brought for her on his last visit to the Old Manor against the wall in the library and it prompted a soul-searching letter of shame, a sentiment that didn't visit him often.

```
22.4.82
Lono Skybridge
. . . Apres moi, la Deluge . . .
Dear Ralph
I have been so crazy with grief and shame
```

about what I did to Sadie's radio that nothing
would console me for nine weeks.

Why did I do it?

I don't know, Ralph. I must have been out
of my head. I was crazy from deep culture
shock . . . & I am SORRY SORRY SORRY

23.4.82

Ralph

FYI

OK. The joke's over now. I had a good
time writing the letter & I hope you enjoyed
it - but don't expect anything else for a
while. I just got a call from my lawyer,
telling me to 'start thinking about packing
up . . .' in order to vacate the Owl Farm
on June I, unless I can come up with a
cashier's check for $132,000.00 to give Sandy
at that time.

Which is out of the question, of course -
so my focus will change for a while, back to
the dark lane - lawyers, money, etc. Plus
moving out & finding a new place to live.

Bad business.

Sandy has been touring for most of the
year, & not even Juan has any idea what she
plans to do when she gets back - which will
happen just about the time I was planning to
hit the wall in Kona.

So I'll give it a bit more thought & let
you know. Ian Ballantine will be out here on
Tuesday with a dummy & maybe we can do something constructive with LONO . . .

Or maybe not. Like I said - the joke is
over. Fuck these people. I should have burned
this place a long time ago.

Good Luck. HST

```
22.4.82
Dear Ralph
Guess where I am going next week . . .
You already know, don't you?
Yes, because I must. For reasons you know
in your heart. Help! HST
```

Then, Hunter was invited back to Hawaii to enter the 1982
Fishing Tournament:

```
The call came last night. I heard the music
& my brain went limp with joy.
YES! YES! I shouted. It must happen!
No problem, He said, & His voice was the
music. Laila cried & called me a faggot — but
nothing matters now, Ralph; I'm back home, to
my people.
And this time He offered me the Chair —
no more of this chicken shit squabbling about
who can catch fish & who can't.
They know, Ralph. Even Alex & Norwood caved
in.
```

They need me — and so does the Tournament which has been moved up to May 5-6-7, in a desperate attempt to crank up some semblance of business, which has gone from bad to worse. Even the rats on Quinn's patio are starving.

So I think they should pay my expenses this time — & I think they probably will.

So you understand, I think, that I really have no choice. It's the only way to finish this wretched thing.

And if you ever mention the idea of us working together again, I'll kick yr. nuts completely off yr. body.

Which reminds me . . . yes . . . I will have to keep an eye on that Black Belt drunkard that I kicked in the balls last time. He will be stalking me . . .

Maybe I should take Murray over with me — & give him the Gaffer's job, for $500 a day.

That would cover my incidental expenses & also teach him a lesson.

But not you, Ralph. Not this time. Maybe I'll give you a ring from that pay-phone in the breezeway of the Kona Inn bar — or on a radio/telephone hook up from a boat far out to sea.

They laughed at Thomas Edison.

And they murdered Capt. Cook, then fed his heart to dogs.

Remember that feel of the boat rolling under your feet? The scream of the reel when a big one hits the hook?

The fine meaty sandwiches? The squawk of the Radio?

The shame? The heat? The black flag? The wild whores dancing on the dock?

Yes. You remember these things. They humm in your blood, as they will humm in mine next week.

Hot damn! And good luck in the Falklands.
[...]

That fleet will never come back, Ralph.
Sell yr. castle to Martin & get out while
there's still time.

This is it, for the Empire. Remember Capt.
Cook.

Flee to Kona, Ralph. Change yr. voice with
a tongue-clip and dye yr. hair pink for the
planeride.

There are still two (2) open slots on Team
200. So far it's only Alec, Norwood, Capt.
Steve and me.

That's $30,000, split 4 ways – which is
more money than I'll ever make in England.

Or you either, I suspect. A lot of people
are going to be filing legal actions against
you pretty soon. They'll seize all yr. money
& freeze all yr. income & put you out on the
road with nothing but a wheelbarrow full of
gimcracks for a tombstone.

[...]

Right. And that's about it for now, Ralph.
You handle my theatrical publicity in London
& I'll handle our action in the City of Refuge
– which is also on the market, for $10K in
front & a huge balloon payment in nine years.

But we can beat that one, Ralph – just
like we beat all the others. Just send me a
cashier's check for $10K & I'll take care of
your yr. debt to Bantam.

It never really happened anyway. They're
all gone now – fired, crazy or dead.

The only one left is our new editor, Ian
Ballantine who has a whole new plan for the
book – 100 pages for 5.00, with no inside
color at all, & a $29,000 phone bill to cover
before we make any profits.

It's all politics from now on, Ralph — & that's bad news for you, once again. By the time we get out from under this dirtbag, you'll be lucky to get a job tending the rat-traps at Quinns.

Yr. wife will be bagging groceries at the Jap Supermarket in Kaillua & yr. children will be spread out all over the Pacific working as indentured servants on atolls like Guam & Samoa.

The Koreans will get Sadie. Wilbur has already arranged it, with a man named Kwan who runs a lounge in Waikiki.

Tell yr. agent to call me about the money, & papers for the children. With any luck at all, we can pay off the Bantam debt by Xmas & have money for drugs left over.

Don't worry, Ralph. I know what I'm doing — and if you have any doubts, come down to the dock in front of the King Kam Hotel around 4.00 pm on May 7, when they hang up the winning fish.

That's me, Ralph. You had it right all along. I was a fool from the start & now I'm going back for more.

But this time I'm rich, for reasons you'll never know — & I'll be a lot richer when this world-class fishing contest is over.

My share (as 'the Angler') will be around $10,000, tax-free, & I've already signed papers consigning the money to a sculptor from Kyoto who works in bronze. He has designed a huge & imposing statue of Fatty Arbuckle to guard the entrance to the City of Refuge.

I don't know much about art, Ralph — but I know what I like, & this one is definitely a right & stylish thing to do.

Thank god we finally turned yr. rotten

little scheme into something decent, or at
least beautiful . . . & that's all either one
of us really wanted in the first place.

We're good people, Ralph & we do good things
for the world.

Cazart. H.

27.4.82
Dear Hunter & Laila
Waiting for Godot finally pays off. The scum
left at the high water mark dries out in the
sun — Scurvy is the worst thing a man can get
at sea — the fish have gone South — our cesspit
is whole again — we first heard about the
Falklands crisis on the radio — so we sold
it and brought newspapers — Bernard asks me
if I've heard from the great man — Bernard
makes a money making suggestion — Woody Creek
is saved — but not for you — for every time
there is a season — the natives are walking
down my drive and they want to smash my
machinery — but it seemed a good idea at the
time — the media waits without, for word of
your coming — Metnick knows — the doomed generation have nothing to lose — my children are
sailing on the next tide — Theo makes his
first footstool and lives like a King — I'll
handle the publicity if you win the tournament — my portrait of you will be on the TV
Times for our programme — this country has
not finished with you yet, old Sport — for
England & St George — it's back to the old
country — America was never for you — too big
— and Hawaii is too small — but you never
learn — your kind never did — you're a red
neck & what's more there are boils on it —
your jaundiced view of life is keeping the
babies awake — Let them sleep — their time

will come when it is not your job or mine to tell them what is wrong – don't be a spoilsport. Play 'the Farmer's in his den' like you did in Hawaii but this time don't frighten the children – and I thoroughly enjoyed your letter – so did Martin – and Anna.

Our fleet will come back. Make no mistake. I think we have been pushed so far as we can go & we are in one of those 'scrapes'.

Of course I am Welsh so I am only an onlooker who waits on the sidelines – then goes into the scrap metal business. Life must go on even in the 80's and bear my drawing in mind viz 'Whoopee! We've won!!!' It is difficult to see 1992 with a straight face. It doesn't sound like a year at all. Just a combination number to a safe.

By all means take Murray with you on the tournament but get a camera on board. Then you wouldn't have the pain of trying to make the film after, at Universal Studios in a plastic tank. Do it for real this time and pay him as an extra. Then you'll be RICH!!

Since I will be over here minding the shop I won't take up your veiled offer to form the crew. I would only hold you back and look silly. I don't want to look silly. I'll do a good job on your publicity in England. I hate boats. You know this. I get sick and I whimper in my cups. I don't think you ever realized the brave faces I have put on in the past to look like a man among men. Real he-men who grab the hairs on their very own chests & tear them out by the handful.

I have avoided drawing those people for fear of reprisals. They frighten me with their own fear of themselves and they can't speak properly because their big chests push their

chins up under their noses. And what's worse
- we're ruled by those men, and it worries
me that you may be one of them - against your
will - a flower in a macho prison - when you
could be like me - a sensitive faggot. Woe,
and God help us.

No! Hunter. Catch the fish for freedom &
for me - and Laila. And Anna too - and Sadie.
And Ronald Reagan too if you must. I know you
are proud of your country.

[. . .]

I don't like drugs, Hunter, as you know
and neither does my son, Theo, anymore - but
I will defend your right to absorb them.

It is a pity that you have engaged a
sculptor who works in bronze. I have just
begun to work in iron and you probably never
realized it.

Ian Ballantine will do the right thing if
only for my sake. He likes me. He does books
for artists. People who think & feel & sing.
If it is 100 pages in black & white for $500
then he is right and so were we all along
and working with you is a pleasure which you
point out in your letter.

I have an idea about the Falkland Islands
which I'll talk to you about when we meet.
But let's not be casual about this one - let's
do it with Draconian zeal. Let's bear in mind
that we are defending the Southern edge of
Great Britain.

This idea I have is a musical called South
Atlantic! And it opens with all the sailors
standing on the deck of HMS Hermes or whatever singing: 'There's nothing like a dance'
etc and in the background my picture of Maggie
Thatcher in a porno movie on a video back
projection. The sailors jerk off with the audi-

ence who become part of the show - when they
realize they are no longer the audience but
only semen fodder. Like it? I've been working
on it for weeks and I want Andrew Lloyd Webber
to do the music if you will do the lyrics.

We have got to go down singing. It's the
only way. It's a blockbuster. It could cause
a Third World War! People would flock to see
it. Then we'd be RICH! We would clean up on
the first night and there wouldn't need to be
a second.

Ring me from Kailua.

Best Ralph.

11.5.82

All in all you left a good impression.
People think you're a wow - and in spite of
my dark protestations and moral outrage, I
can convince no one that you are an absolute
shit whose veins run green with lizards bile
and pure cocaine.

Hell's fire will take you - but not yet.
There is work to be done and that's what makes
it all worthwhile. It's the winning smile that
softens every heart.

I have heard nothing yet about the Lono
book save for a grape vine message via Abner
who heard through some deep throat that Bantam
definitely want to publish in the fall & now
that they've seen the pictures & the 105 pages
they won't let go. I also heard Alan has been
retired early and has gone to work for Metnik.

My son Theo was sorry to have missed you.
Angry even. You're a weak father.

The Ritz article was awful and no good
will come of it. I believe Bernard is going
to send you a copy of it but I think you
would be better not seeing it at all. Because

of it, two shady youths came visiting. While
one of them was sick down our toilet the other
asked me impertinent questions about how much
I earned. Then I engineered a phone call
telling me that I had to go out and drive
them to the station where I presume they eventually got a train back to London.

Apart from mentioning my disappointment
about you not enjoying yourself here and crying
about it through the length and breadth of
our halls of residence, we here at the Old
Manor hope you are relieved to be back where
you belong.

I was genuinely sorry to hear about John
Belushi. He was obviously on the crest of some
wonderful wave and it was certainly no time
to be popping off - any more than it was for
our friend Michael Dempsey.

Keep swerving, both of you and have some
sushi for me.

Best,
Ralph

31.5.82
Just a line to say hello and tell you how
the war over here is going.

By the time you get this letter the war
will be over & we will be prisoners.

Don't worry - & whatever you do don't panic.

It is not ours to reason why we are not
the aggressors. We are islanders who try to
mind our own business and lead our lives in
peace & tranquillity.

As you know, our boys have been out there
with our unequivocal support putting right
what is wrong & bringing home the bacon as
we say over here - or sending home the bacon
as they say over there.

We over here are not over there but our
hearts go with them and the pope is over here.

He came over here because he did not want
to go over there — but he may well go over
there.

Don't believe for one moment that over
there is where he wants to be. Because he
said so.

You have probably been fed huge daily rations
of propaganda. Lies about us. DO NOT BELIEVE
them. We are OK and we don't mean any harm.

We love people and most of all we love our
neighbours. Don't listen to the evil outpourings of a decadent race of mutants. What they
have to say will only confirm what I am trying
to deny: We are innocent and our navy was
only on a routine exercise. Our boys only
wanted to be in the Lamb and Flag public house
playing darts.

Do you really believe that we would send
a whole task force out there half way around
the world to knock out of the skies Sea Harrier
jets 'owned' by the Argentines, supplied by
us!!!, which have not yet been paid for?

What kind of fools do you take us to be?
What stupid idea gives you the presumption to
imagine such an assumption?

Why do you mock our ways? And why is it
in the name of Haig's peace mission you leave
our shores faster than an EXORCET [sic] missile?

Have you no shame?? Your ancestors are
Scottish, aren't they? — or are they Irishmen????

They can't be Welsh or you would stare at
life more than you do and wonder why.

Thompson is not a Welsh name anyway so we
are at least blameless of anything you might
have intimated.

Never mind, Scotsmen have big hearts.

I enclose a few pieces of ephemera for your personal collection of grandchildren items. Things to point to with a twisted stick from your sickbed with pride.

Anna and I are off to France for two weeks. We will write to you from there.

I read something the other day about Leonardo da Vinci. Some people are nothing more than receptacles for the passage of food — and I thought Hot Damn!! That's me!!

And they laughed at Leonardo.

God Bless. Ralph.

21.8.82

I began this letter when Ian was here on 21. Aug 82. Then I had to come to Milan — read on.

Dear Hunter and Laila,

It's like learning that your sushi partner eats cooked fish!!!

I heard about your outrageous new contract from Ian Ballantine only yesterday & it hit me like a jackhammer — you wanted to kill off this awful Hawaiian nightmare because you could now afford to pay back the advance — whatever Sandy got — you swine!!

What sickens me most is that all of your fans probably can't read anyway & prefer to look at pictures and yet YOU get the big one.

So that's why you like fishing — you don't like fish at all — not even cooked fish — you like money — lots of it & you don't care how you get it!!!

I had decided to visit Milan to see the *Last Supper* at the Santa Maria delle Grazie for *I, Leonardo*. More gibberish as far as Hunter was concerned. At this point *The Curse of Lono* was still not complete . . . but it was getting there.

Just for something to do, I wrote a letter to Hunter from Italy.

Milan Sunday 28 Aug 82

Ian Ballantine is a gentleman & I always get on with gentlemen because above all they speak in decent English.

I have read all your new pieces & I love them but you must remember I covered all eventualities you could dream up save one - the dog with red fleas.

The drawing is now done - one of the best for humour & spontaneity. Some things are funny & some things are sick. This scene is funnier than the hotel room in Las Vegas & not quite as funny as the letter to my son Theo.

Never mind. The important thing is - you always can turn a phrase like 'gnawing on balls of twine' and now - get on with your great NOVEL. 'The SILK ROAD' - I think your graft is over for Hawaii.

It's an odd story - hardly that, but certainly a record of a strange convoluted journey that we took - for good or ill. You on your boat and me on nothing but a ticket to ride.

I'm in Milan at the moment.

There is hardly a stone left of that period and I wander the streets like the buggering fool I am desperate for a break.

People finger me for money and in Paris I made friends with a bag woman. I bought her a drink and I bought her a meal on a plastic card & acted crazy in the restaurant - of course, crawling under tables while people tried to eat spaghetti & pretend I wasn't there.

Never mind the bag woman - I was the embarrassment, the person they least wanted

to meet over a quiet & romantic meal by
candlelight.

There is a craziness even you don't
understand & I have found the key. They are
moments I cannot repeat, neither would I
want to.

When we left the restaurant even the bag
woman said. 'I'll see you again on the street.
I will walk in the other direction.' But nevertheless for all that she likes me because I
was kind to her. Kindness is a devastating
force - just like cruelty.

She was not 'coyote ugly' - here she is -
[the Parisian bag woman] but she needed a shave.
She disappeared into the night as gently as
she came out of it.

So, you see, I am still studying human
nature - for my work & for the good of mankind.

I believe you have cut your lawn at Owl
Farm - but I also believe you are not too
good at it.

I must away now to the Sforza Castello &
also see the Last Supper, for tomorrow I leave
for England & my own Castello in Kent where
you lived in peace and contentment for such
a brief period in the depth of winter.

The sushi in Milan is exquisite but expensive.

God bless.

Ralph.

By September it had come to this:

9.9.82

Okay, Ralph.

It's 50/50 time - & I can't afford to dump
this book (as you know) as much as I would
like to.

By my lights, in fact, Ian [Ballantine] already has a 200 page manuscript, which I gather you've seen & worked from (inre: Red Fleas) [a reference to a dog infested with red fleas that he claimed he had bought for my daughter, Sadie].

We are already over the limit(s) defined in the contract ('Just sign it' - remember that advice?) And the only thing that kept me from sending at least 100 finished pages 3 weeks ago was my slow rising sense of hatred for the opening chapter which only looked passable in England, when my mind was somewhere else.

I knew in my heart it was lame but nobody would say it - until Juan arrived & said the problem with The Blue Arm (which I'd been trying to cure with a malignant dose of insects) was the fact that it was dumb & wrong from the start.

Which was (& is) true. Bad writing, worse dialogue - a disastrous opening 'chapter,' & a curse on my head forever.

Ian is doubtless a fine old gentleman - but he is a publisher, not an editor, and when I finally squeezed an 'editorial' opinion out of him - in a moment of real stress, on a bad afternoon in the kitchen - he hit quickly and sternly on two scenes: 1) the memo to you about looking for Pele & picking up female hitch hikers & taking them out to the Kona Surf parking lot . . . and 2) The Ending, which also mentions tits.

Ian doesn't want Tits in the book - but I do & we will have them.

'Just finish it.'

That's the only advice I get. 'Never mind what it says - just spit out a final draft. Finish it.'

Which gets us back to 50/50: we are co-authors on this book, Ralph, and 'S' comes before 'T' in the alphabet.

I have spent almost two years trying to tell Bantam that I will not under any circumstances allow this thing to be published and/or marketed as my 'next book.'

I have my own book to write - & LONO is not it. I can send you whatever you need repeat need to wrap around your art & satisfy the contract - but it ain't 'my next book.'

It's your next book, Ralph. The art should be the selling point - not the bogus & essentially treacherous use of my name.

Ian is already scaling the art-text ratio down to about 20/80 - (totally contrary to the sick & potentially fatal (to me) numbers we agreed on) - and our original notion of a quick 50/50 mix of art and text now seems like a stupid joke.

They never took that seriously in the first place - & if they persist we will have a problem on our hands when they send the bugger to press.

And we don't need that. It will be hard enough to explain without blaming all on me . . . & making me deny it.

We don't want that, Ralph. This thing has already put serious strains on our friendship, & if it festers much longer I might not survive it, as a writer.

This is the Curse of Lono. I was right from the start - & I never even got paid for it.

Ian said 'There is no deadline' for this thing, but for me there is, and it's now.

If it's our book I can send you the first 100 pages at once (including a re-write of The Blue Arm) - but if it's my book, the joke

is over. We will have to re-adjust the numbers
in the contract in a very meaningful way, to
fit Ian's format & sales pitch.

So . . . yes . . . here is that nut: You
should take my first 100 pages & start lashing
the book together, as if it were yours — which
it is, on the payoff. You have much to gain
on this book, & nothing to lose.

My position is exactly the opposite. All
I have to gain is . . . well, let's just say
it's not enough, Ralph. I don't mind helping
you out — but not if I have to die for it.

As far as I'm concerned, we already have
more than the original contract ever called
for — it was supposed to be 'Postcard from a
Wrong Vacation' — not 'Dr Thompson rides again.'

Why would it ever occur to me to give away
50% of 'my next book'?

So let's just start on the bugger. I'll send
you 100 pages with blank pages for art, in
sequence, & other graphic options in the way
of quotes & sub-heads. Nothing we haven't already
discussed — but not an absolutely final draft
either. I'm ready to try it, if you are — 50/50
down the line. Let me know. Thanx. HST.

It was at this point that I knew Hunter had found an
entirely new horizon that didn't need my help. He had found
a new passport.

So I wrote to Ian on 15 September 1982:

15.9.82
Dear Ian
I have just come off the phone after
spending the last 2 hours of conversation with
Hunter regarding the current situation with
the state the book is in and the form it
should take in order that no one should be

in any doubt about what sort of book this is, whose book it is and what can be done to ensure that the end result reflects as closely as possible the nature of our collaboration.

Before continuing, I feel obliged to point out that our talk on the telephone was in no way a social call and was the result of a cry for help by letter from Hunter. It was a letter that disturbed me intensely and that caused me to phone rather than try to write — which I consider to be in the nature of an editorial call & hopefully a call for which I can be reimbursed. I'm banking on Putnams having a petty cash box for such occasions.

Believe me, it is cheap at the price for I now have a much clearer idea of just what continuously prevents Hunter from letting go of his pages in a form that he can find acceptable.

Primarily, this book is not a NOVEL or even vaguely in novel form.

It is, and will always remain a collection of disconnected notes written in past & present — the sense of being there — which is Hunter's great strength — and the sense of reflection which does not necessarily form a connecting narrative.

That is, it does not necessarily follow that because Ackerman is mentioned in Chapter I page?, for instance, he should ever be referred to again. It may sound crazy but it does not have to be.

These are recollections and not intended to form the basis of some plot or similar device authors use to tell a story. They merely happened and some incidents are worthy of re-telling.

There is no reason for threads to be picked up & woven back with the 'story' any more than pictures should relate directly to any of Hunter's text.

This was an experiment from the start - and I believe it should remain so & most important it should feel like fun rather than an examination for a degree in English literature or a degree in art appreciation for that matter.

Well, that is for you as an editor, publisher & manufacturer of the end result and me as 50/50 partner to a mind which works best at scatological levels with its own built-in humorous intelligence, and the inspiration for some of my best drawings - and I can give no greater praise than that!! - must treat the material with the same respect as a great comedian would choose a certain nuance here, an inflection there, to convey the spontaneity in a joke as though he were telling it for the very first time.

And it is a joke and something which Hunter is dangerously near forgetting because of the apparently wooden approach a corporate giant like Bantam Books is imposing on this project. Albeit well-meaning.

I fear the worst if the said manuscript is considered as a manuscript like a block of type.

It is not a block of type but a glimpse into a private collection of observations and should be treated as such; i.e. in various type-faces but without the usual dictatorial grid which would inevitably place it in novel form.

I believe I hinted at this at the meetings we had, but it is a great pity that I had not spoken to Hunter before we met as I could have explained myself a little better and conveyed most of Hunter's real fears about this book.

Here is a list of fears — and real or
otherwise, they exist for Hunter anyway and
I personally would prefer that you heed them
now rather than later.

1. It is not his next book!
2. It is not a novel.
3. It is not understood that notes, letters,
dialogue, grocery bills, connecting prose are
taken down and used in evidence against him
as nothing more than a publishing scam to buy
his name cheap and make huge profits — even
for him — or me.
4. That he has no control from the moment
his words leave his hands.
5. That there will be no opportunity to
reconsider a phrase here or a change there in
mock-up form & even before mechanicals.
6. That once the text in whatever form
reaches you it will be 'peddled' mercilessly
on his name — and not mine.
7. That while I can be justly proud of
my contribution, he does not like his effort
. . . particularly because it will be misconstrued as a 'serious novel' & therefore
criticized as such — or as a piece of writing
of which Hunter is justly proud.

Well, he may be yet, if the material is
treated in the spur of the moment which
is how it was all done, though lately it
appears that it has become a burden beyond
its importance.

I never had any doubt that it was a fun
book and if there is pain in it then that
can only be a bonus — for our collaborations
have always contained a hint of that.

I think it most certainly should be made
clear that experiments like this should only
be attempted by those who are completely aware

of - or even know - what they are doing &
by saying that I am appealing to you to
consider the statement as a cornerstone on
which to build a decent, honest and humorous
promotional campaign.

In a sentence - to promote the book as a
monumental fuck up. It will sell a million -
in Hawaii alone.

God bless.

Ralph Steadman

For a while I did not hear from Hunter. But, in March
1983 he wrote to me:

Dear Ralph

Sorry to be out of touch for so long.
But it seems like I've spent the last 2 or
3 years trying to get way from that Goddamn
wretched LONO book. It haunts me like a
cancer in the nuts & I would kill it if I
could, but the shitwheels are already in
motion . . .

I was right the first time. It was a
deathtrip all along like that stinking movie
that spawned it.

Ah well - the art's not bad, & I guess we
had a few cheap laughs along the way.

Scum. All along - nothing but scum. And
in the end it will come out to the same old
2 and a half %.

Res ipsa Loquitor.

HST

11.5.83

Dear Laila and Hunter,

Relax. I'm in control. I've had the whole
of the Bantam Art Dept. in a frenzy of upheaval.

Under my careful direction great changes

are being wrought & you will be proud to know me. It is all in hand. The book will fizz with unadulterated brilliance. You will be thrilled with the result & the book will sell millions. It should now be called The CURES OF LONO.

I have a native understanding with these people.

You can sit back & watch now. The battle is won. The light has been seen. The book will shine like a piece of literary sculpture and it will win prizes.

I asked myself - what would Leonardo have done - & the answer came back - use Welsh diplomacy & all will fall into place. The next time we see a dummy book all will be well.

Concerning my other book called I Leonardo I have been in close consultation all along with the designer who is the firm's art director anyway & that too will be a masterpiece of intuitive collaboration.

I have written the text in the first person in a 16th century style of English which I believe you people took with you to the colonies & which accounts for your strange guttural accents.

If that were not enough I have filled it with the most lustrous renaissance style pictures. The next book we do together, Holiday in Belfast, or whatever it is, must be cleared in the beginning so that contractually we must be consulted at every stage.

Best, Ralph.

And that was that.

Hunter was right. It was all gibberish, his as well as mine. But the letters were enjoyable to read and place us at a time and place when things were moving too fast to remember.

1996. The Lotus Club, New York. The *Rolling Stone* twenty-fifth anniversary party to celebrate *Fear and Loathing in Las Vegas*, first published in the magazine in 1971.

Laila Nabulsi, Johnny Depp, Ralph and Kate Moss.

Hunter and Aspen friend, Cilla Hyams.

The Four Seasons Hotel, New York. Laila comforts a dead-beat Hunter after the celebration.

Ralph, Jann Wenner, Janie Wenner and Hunter.

1995. Hunter and Ralph cutting the cake to celebrate the twenty-fifth anniversary of the birth of Gonzo in the kitchen at Owl Farm, Woody Creek.

Studies of Hunter in his fighting chair at his kitchen counter, Owl Farm.

2004. Scenes in the kitchen at Owl Farm
and discussions about the unfinished
Polo is My Life with Belinda, the four-eyed
horse over the piano.

**GUN SHOTS
WILL BE HEARD
DURING THIS
PERFORMANCE**

he final mystery is
oneself.... Who can calculate
the orbit of his own soul?"

OSCAR WILDE

Virginia Thompson, Hunter's mother, aged ninety-two, in her room at her care home, Louisville, Kentucky, surrounded by her family photos and paintings during Ralph's visit in 1997.

2004. Ralph working on the title
Vote Naked and Die at George Stranahan's
ranch house.

Oxygen for Ralph.

A spliff for the pig at the Woody Creek
Tavern.

Kangeroo court poster at Owl Farm.

Hunter with good friend Sheriff Bob
Braudis.

The cheque that Ralph gave to Hunter for
signing the limited edition of his last book,
Fire in the Nuts.

2005. The Jerome Hotel, Aspen. Johnny Depp signs fist monument print based on the original drawing created by Ralph in 1977 and printed by Joe Petro, to be sold in aid of the Hunter S. Thompson estate.

Bill Murray shows photos of his early acting career.

Ralph impersonating Hunter.

2005. Ralph with Anita Thompson, Hunter's wife, just after the March memorial.

Ralph with Laila Nabulsi in Owl Farm kitchen.

1979. Drawing of Hunter from the back jacket of *The Great Shark Hunt*.

DRAWING © 1979 BY RALPH STEADMAN

BANG!

Hunter S. Thompson

2005.The fist monument, Woody Creek. First conceived in 1977 in a West Hollywood funeral parlour which Hunter asked Ralph to design. Joe Petro, Anna and Ralph Steadman seated beneath the realized monument, funded by Johnny Depp and activated to send Hunter's ashes into the firmament on 20 August 2005.

Even slightly out of order they anchor a time to itself and me to a period when it was all such a wild game.

It was following these titanic struggles to make all seem reasonable that I took on another Goliath of a project somewhere around 1986. I had been so embroiled in gods, Hunter and God, Leonardo and God, that I determined to do a book about God and why he/she/it was such a vindictive old bastard and what the hell it could have to do with me. So I began *The Big I Am*.

'Ah, Ralph! I see you have written some more of your psycho-gibberish. No one is going to believe you,' Hunter would say. 'You will end your days wandering the streets, grabbing people by their coat collars and dribbling your new-found religion on their shoes. No one will buy it, Ralph, and you will be a freak like John the Baptist in the Nevada desert, living off cactus and juniper berries and sobbing like a queer vicar trying to start a new religion on the fringes of Area 51 – right out there on the edge. But don't worry, Ralph! I believe you. You're right! There are UFOs everywhere but it's too weird for most people. I always knew that behind that gentle exterior there beats the black heart of a demon on a mission, seeing through a glass darkly.'

I can hear him saying such things to me in his Kentucky accent, his staccato mumble, his well-honed glossary of familiar mantras, his litany of mean humour with just the right amount of mockery, thus avoiding having to say: 'Good grief, Ralph! That's good.'

Comparing his later letters to the early correspondence, the change in tone was palpable. For a while there, towards the end of the eighties, we were both doing quite different things and we both knew it. My drawings for George Orwell's *Animal Farm* from 1985 drew silence. Or, he would say something like: 'Just another fish wrap journalist who got lucky.' That would be a familiar and harsh judgement about someone who had managed to rise out of that trench warfare and write some real stories. His contempt masked an admiration that would be seen as a weakness and that would never do. All his heroes like Joseph Conrad, Ernest Hemingway and William Faulkner wrote proper stories and then there was Hunter, this magnificent outlaw, with jangling silver spurs on a pair of Converse Low basketball sneakers, whose prose style was peerless, but whose ability to write a novel eluded him to the end. He was his own best story.

But I haven't finished with him just yet. I need to mock him and beat on him like a lost cockerel, just like he mocked me.

THE YEAR OF WINE
1987

*Drunk on the road to Denver . . . Hell's Angels at the
bookstand . . . Today's pig is tomorrow's bacon . . .
Meeting royalty . . . Whisky not wine . . . Polo is not
my life . . . Wedding bells and no smoking at the inn*

Nineteen eighty-seven was the year of wine and new directions and the year I stopped drawing politicians.

Hunter was disgusted that I had taken such a route. He
said to me – 'Grapes! Stupid little grapes!' his voice full of
contempt. 'Fruit! Politics was beneath you,' he said, 'so you
stooped to worship grapes.' After that he heaped it on and
imagined me guzzling whisky, grapes and using hired help –
he dubbed it a nasty fermentation. No wonder I was sick. It
was as though a long-buried Southern gentleman had risen
up from a dark, dismal hole, brandishing a horsewhip and was
going to lay into me for breaking the code of the proud sons
of the Kentucky Pioneers. I was his whipping boy again. I was
wrong as always and in his heart he was enjoying the opportunity to thrash my stupidity with insults. Although he drank
wine on occasions, in copious gulps, it was jammed in between
the grapefruits, Bloody Marys and Chivas Regal.

He was the messiest, most piggish, food lout I have ever
dined with. The plates of unfinished room service orders I
have seen over the years in four-star hotels were battlefields
with mountains of cold, melted cheese, torn hamburgers, deci-
mated salads and tomato ketchup – a massacre on a food trolley.
They were battlefield exhibits of the carnage he inflicted on
all food and convinced me that he wasn't interested so long as
something filled a space in his stomach.

But that I chose wine to dwell on and to draw pictures for

a chain of English wine merchants called Oddbins made me worse than Hitler – the Reichsführer super-Nazi, he added, as though he was pouring some extra insulting herbs and condiments into his hellish concoction. To this elegant thug of a wordsmith, the name was never enough and he ladled superlatives into the bubbling sauce like an eager apprentice in the devil's kitchen. Even then he would scrape the last slops from the bottom of the pan and, with a flourish, flick them onto me to make sure I got the full taste of his streaming insult.

He treated my visits to wineries all over the world as a total sell-out, a binge of self-indulgence that I was born to accept like a cheap pimp feeding freely off the fat of the land. When I started writing as well as drawing about it, he hammered home his contempt with warnings that writing was the end of the line for me. I became a serious geek who could not function in anything Gonzotic again.

In the August of 1987 Hunter promised the Edinburgh Festival that he would come over for the occasion and address an audience which would fill a big marquee and I would be there to welcome him and introduce him to the fans. He could look forward to a weekend of grouse-shooting and a spot of golf at St Andrews. I maintained phone contact with him right up to the moment he left Owl Farm and climbed into the taxi which was to take him down Route 70 East to Denver and the airport. He was actually on his way and I was very happy. I told everyone he was on his way. Then everything went horribly wrong.

I got a call from a restaurant phone booth somewhere around Vail. 'Ralph! I've got some dumb, drunken freak for a cab driver! He's drunk, Ralph! He's been drinking my vodka and there's a gang of hippies lying around on a slope and they are trying to levitate. It's weird. I can't make sense of it. We still have thirty miles to go, too,' which put them somewhere around Georgetown, if in fact he had even left Owl Farm.

'There's going to be a huge crowd here – and the Queen of England is expecting you at Balmoral,' I added hopefully. It had been mentioned by a newspaper editor (Simon Kelner, working for *The Observer* at the time) who was trying to fly a story on it. Meeting royalty was a big part of the enticing equation.

'Ring me when you have more news,' I said. 'I'm banking on your appearance. Don't let me down!'

'I won't, Ralph. I've got a plane to catch.'

Two or three calls later he was still buggering about at 6,000 feet up in the mountains, trying to steer the car for the drunken taxi driver from the passenger seat and keep his foot on the brake pedal, just in case, while plying the guy with more vodka.

'What kind of vodka is he drinking?' I asked as though that would make any difference.

'Cheap!' came the reply, followed by a 'Ho! Ho!! Ho!!!' There was no way he was going to catch anything but altitude sickness. Call it a style. Anticipation, exhilaration, inebriation and exasperation . . . but I knew it wasn't going to happen. I tried to prepare people for the worst because I knew in my heart that there had been an almighty fuck-up.

'That could be a problem,' said the organizer not long before the event, 'because the Spiegeltent is going to be heaving with Hell's Angels. They have ridden up from London to meet him.' Later, around the Georgian Charlotte Square Gardens, where the Book Festival had been set up, people had been arriving in droves, already half-dosed to distraction, having paced themselves all night for the event.

Out in front of the Roxburghe Hotel it looked like a Trafalgar Square Peace Rally preparing for war. It was a scene of quintessential 'Bad Craziness' and I would have to go out there and say: 'Hey, you guys. It's great of you to turn up, but instead of him, here is a spot of light relief in the form of nine wayward nuns in black skin-tight rubber habits singing "Walk on the Wild Side".' It would have calmed the crazed throng down for about fifteen seconds before I would have been grabbed and lynched like a rag doll. I refused to go out to try anything resembling an apology. I left it to a scrap of paper that Hunter had taped to one of his cupboard doors over the sink in his kitchen. He had written it during the night when I was at Owl Farm grappling with some final touches to the

Lono debacle years earlier: '. . . To show man the best that is in him; not the most appealing or the most amusing or even the most realistic — but the best, which is rare and common and understood by all of us in all our different ways . . . to include all the others — the meanest, the cheapest, the most cowardly — as a background and a foreground for something better . . . to dig in the old scum that covers us all and find something that might be a tool for a man who would use it to fashion his self-respect in a world where all those tools are buried or broken or illegal . . . and finally to tell it as it is, trying to see it all and especially the best, for to miss that part is to shovel shit on men who were born in quicksand and find no novelty in the heave and smell of doom.'

THE 90s

'The joke's over, Ralph. You've sucked on my
back long enough.'

OWN GOALS
1990–94

Legal Defense Fund ... Plagues & moonflowers ...
Friendship adrift ... Bleeding England ...
Hunter's retreat

Faxes kept coming fast, before, during and after this period.

30.12.89
Ralph
Why? Does yr FAX machine send everything
in triplicate??? Thrice. 3 copies each???
Weird . . .
What are you up to?
I have a book that you might want to do
some drawings for — chronicle my degeneration
thru five decades . . .

Hunter.
My fax machine is perfect — it was merely
telling me that somebody was using an inordinately large amount of FAX paper all at once
and maybe there was a maniac on the other end
— to press the red warning button & alert the
police.
OK. Must retire to the pub to drink and
digest what I have just read which appears to
be fiction or a real nightmare — or something, certainly racist! I will FAX scribbles
through shortly.
* visual ideas OK? Jann is offering PEANUTS!

One fax just before the new decade was particularly hilarious.

25.12.89

Merry Christmas

Ho, Ho, Ho,

I killed a dog yesterday - a big hound, menacing the peacocks - with a sawed-off 12 gauge riot gun, but we had trouble getting rid of the body, which was heavy & bleeding profusely.

I grabbed an old Red Santa Claus suit.

It was like the Too Much Fun Club: white Xmas, whiskey, fire, presents heaped under the tree.

But the children knew the big stuff was not there yet. They were waiting for Santa Claus.

We also waited after lashing the body into the red suit with heavy twine, we put a Santa Claus hat on it & went to the Tavern for dinner.

George was there along with most of his clan. 'The kids are already in bed,' he said. 'They're crazy with excitement. They all sent their Christmas lists to the North Pole, now they're waiting -'

I took it off a nail in the basement (red nylon with white rayon trim) - and wrapped the dog in it, like a shroud.

It was Christmas Eve and we had our own Santa Claus.

I also had a score to settle with Stranahan, who has a house full of relatives for Xmas - including 4 young children, ages 4 to 9 or so.

The house was full of Christmas cheer, egg-nog, laughter, yule log, crackling in the big fireplace & 12 stockings hung from the mantle.

It was long after midnight when we drove

up the road to his house and climbed up on his
roof, moving quietly in the new snow, dragging the
carcass behind us on a rope.

We lowered our bleeding Santa Claus down the
chimney very slowly - almost all the way down to
the smoke ledge above the living room fireplace
where the stockings were hung.

But there it stopped. We were unable to move
it with the rope, so we fled. It was almost down-
hill & there was a smell of slow burning flesh
in the air.

Soon there would be blood dripping into the
fireplace, chunks of burnt red cloth on the hearth,
foul smoke backed into the house, setting off smoke
alarms & waking the children. Ho.ho.ho

Later — some time in April 1990 — he shouted abuse at me
for not raising sufficient funds for his Legal Defense Fund
when he was accused of trying to rip the nipples off a woman
who had come to interview him. It was probably a playful
moment that went weird but he was accused of technical
assault. A gonzo-loving Australian friend, David Langsam,
persuaded the *Guardian* to let him write an article about
Hunter's plight. David gave a fair account of the whole sordid
encounter — well balanced and true, thus helping to raise more
money for the cause. To accompany the article I did a drawing
of a Nazi pig with a moustache which had been sliced into
rashers with the caption: 'Today's Pig is Tomorrow's Bacon',
referring to the wicked law that had put him in jeopardy.
Well, it was funny but I knew nothing of the circumstances
that had led to the incident. 'What about me? What about
me? What about me?' he crowed down his fax machine.

Jesus! You wimpering sot. Your body was made
to be draped over barbed wire - & yr. Wretched
son should be hung up on a festering nail above
the urinals in some hideous public urinal on Queer
Street. Ho ho - I warned you to stay out of the
fast lane, Ralph. Cazar. Doc.

It wasn't particularly friendly, at the time, or even now in retrospect, and once in a while it got too much even for me. Our long twisted friendship faltered.

Then he quoted his beloved constitution – the Fourth Amendment, to be exact, about 'the right of the people to be secure in their persons, houses, pig-sty or wherever else you may want to be safe and carry on doing what you have always done, unrestrained'. He went on:

```
    Why are you mocking me, Ralph? So far you
haven't collected a Pfennig for my Legal
Defense Fund.
    Do you think this is funny? Just another
Gonzo joke?
    No, Ralph - this is deadly serious. They
want to put me in prison for at least 2 yrs.
& the Nixon-Reagan courts are on their side.
    The fat is in the fire.
```

I wrote back sardonically:

```
    Hunter!
    I fully realize the gravity of the situa-
tion. There is no mockery coming from me. That
is in your own mind. I am trying to help. I
have raised £1000 - £250 is on its way to you
since a week ago - £750 is coming from Punch
- the address of the fund is now public. If
all your friends did as much for you as me,
you would be a lucky man. Please don't bitch
at me. I am doing my best. I am a REAL friend
of Hunter S. Thompson. Maybe you just need
RICH friends. I know the spot that you are
in. Things are slow but they ARE on the move.
Please forgive me if I have unwittingly appeared
light hearted. It's just your troops trying
to keep their ends up. OK. Sad. Ralph.
```

6.6.90

This is why you curse me for not communi-
cating with you, Ralph . . . You have a cheap
amateur FAX machine . . . and where is that
goddam 5000 pounds you said you raised for my
legal defense?

My auditors will be speaking with you soon
. . . stealing is ugly, Ralph. Especially when
you steal from the doomed. Send me the 5K
pounds & get rid of that toy/FAX machine. Did
you also steal the money you raised for poor
McCarthy down in Beirut?? So what — eh?

Fuck you, Ralph — you're a monster. You
would cut off my head & eat my brain, if you
thought you could metabolize it.

Get a job, you swine or do something about
this goddamn BBC movie!!! They keep hounding
me for a 'concept'. But I have no concepts,
Ralph . . .

I am a counter puncher, a human lizard with
a small brain. I never knew where I was going,
but I ripped the tits off of everything that
got in my way. By the time they figured me
out it was too late. The dumb die young and
the white Cockroach will outlive us all.

I wrote to him on 7 June 1990, listing the monies I had
raised for his wretched Legal Fund. Much more than I thought!
I turned a blind eye to his thoughtless angst and haggling and
continued to be his faithful Sancho Panza, much against my
wife's remonstrations.

Hunter,
You old charmer, sweet talking all the way
to the bank!
Though you don't need it now here is a
breakdown — and if it hasn't arrived yet blame
the respective accounts departments of the

various journals who have used my work and
offered the following: (I must have been mad!)
     Guardian for use of PIG - £200
     Independent for use of Spirit of Gonzo -
100
     Rolling Stone for use of Hunter's Back -
$3000!
     Punch for use of three drawings - £750
     Bruce Robinson's cheque - £250
     Mine - £250
     Total - that's about $5k
     As for a concept, I have suggested a concept
- The Ballad of Woody Creek - based on the
Robert Service poem, The Land of BEYOND. Like
it? Good! Fax me yes let's go with it and
I'll get the bugger off the ground through
Stephen Trombley who simply needs your collu-
sion - NOT your scrofulous notes to me - and
your schoolboy porn letter didn't go unno-
ticed either, you liceridden degenerate. OK.
Warmest regards. Ralph.

That's nudging five grand raised for him and not one word
of thanks. I barked back at him early one Sunday morning.

     DON'T SHOUT AT ME!! YOU are the one playing
GONZO diplomacy!!! Diplomacy is not your forte.
But maybe that is all part of your hideous
charm. John McCarthy is the one in jail right
now and Ethiopians are dying like flies. I am
involved in projects for both those causes
and I'm involved with YOURS! I KNOW YOUR SITU-
ATION. Don't INVENT scenarios to suit your
vanity. I AM HERE. I WILL DO MY BEST. SO FUCK
OFF TOO - UNTIL YOU CAN BE CIVIL. Ralph.

There was silence until I heard that he had been acquitted
on a technicality. It was all in the past. No more mention of

Defense Funds or anything suggesting gratitude. That was never his style.

Sod it! I thought, 'I'll send him a poem.' It's called 'Friends':

Friendships loom and I run for cover
I fight shy of new friends
Brittle and precious as Dresden china
Dresden burned and China learned nothing

Friendships long gone, built into the past
Friends appear from nowhere and claim my affection
Like winners of raffles

Friendships cemented by clasped hands
Nailed against walls of mindless graffiti
Holding on for better times
Holding on to be heroes

Friendships dissolve walls
Leaving a body hanging over barbed wire like an
    old coat
A disposable reminder

Shining with blood sequins
Polished for the days of reunion
Lives long gone yet reaching out to older friends
Held prisoner in landlocked ideologies

Talons of obligation claw the flesh
What about me?
What about me?
Only yesterday you were my friend

But that was yesterday
When the talons were softly padded in fur gloves
Gentle intentions were only games of chance and
    promise
We played together in the open spaces

Spiked with the nails of older friendships
Sad to say goodbye
Relieved to be alone again
Anaesthetized by the fresh encounter

We wait for news of nothing in particular
Only that the operation was successful
And the friendship survived
The awkwardness gone
We charmed like old pros

Now we can rip the friends apart
Like chicken's legs from carcasses well cooked
And ready to eat
Feelings savaged and garrotted
And left for dead

Friendships loom and I run for cover
I hide inside a memory
Peeping out only to check
Old friends have gone
Needing me like a drug

Around this time I was involved in writing blank verse
and setting pieces to music. I played the guitar and imagined
that music was another part of my creative remit. It was a
time of new direction in a sickening world. In 1988 I was
asked by a marvellous man – a new friend, Richard Gregson-
Williams, who was director of many arts festivals nationwide
– if I would consider being the 1989 Artist in Residence at
the Exeter Arts Festival and, furthermore, would I also consider
writing an oratorio that could be set to music by composer
Richard Harvey and provide a 'background' against which John
Williams, the guitar maestro, could perform.

I had never been asked such an extraordinary thing before.
How would anyone suppose that I could write anything of the
kind? I had no previous evidence of such a thing to suggest
it, but RGW, as Richard was called, assured me that I could

do it. I was deeply flattered and accepted the commission. Who wouldn't? On the train up to London from Kent I had decided that what I wanted to do was something ecological and as happens in moments of serendipity I had read a review of a new book by Margaret Mee, the English artist and traveller, whose book *In Search of the Flowers of Amazon Forests* had just been published.

I was entranced and by the time I met with Richard my mind was made up. I wanted to write a libretto about the exotic parasite, the Moonflower, a cactus that lived off the trees in the Amazon rainforests. Richard was wholehearted about the idea and urged me to go on. But I needed an angle. I had an overall theme – the new millennium that beckoned ten years down the road. I needed a bad guy, an awful threat to this parasitic love object and a symbol of our own terrible appetites.

Anna and I had decided to go somewhere weird for Christmas and we settled on Luxor along the Nile. We embraced the Nile's warmth, dirt and fantasy for nearly three weeks and there I wrote most of the basic plot. For the story, I needed another river and I chose the Amazon, the main artery of an entirely different landscape coursing through the rainforest. The two rivers were and are my opposites; so I then needed a third component. I chose we humans in the form of a plague – our lustful appetites that prevent the earth from flourishing. Man's flawed nature became the Plague Demon which wasn't an alien. That would have been ridiculous. The irony lay in the Moonflower that blossomed only once a year in the moonlight. Margaret Mee had spent years trying to catch one at the moment of blossom and did so once in her paintings using a shaded torch over her painting easel to prevent the plant from thinking it was daylight and close up. Moonflowers give their best shot in the dark. I called it *Plague and the Moonflower*.

The Narrator was Ben Kingsley and the Moonflower was sung angelically by the soprano singer Kym Amps. The Plague Demon (us) fell in love with this vision of beauty as it rampaged across the planet, the part spoken like a Welsh preacher with Hell, Fire and Damnation by Ian Holm, and musically interpreted by a demonic violinist, Richard Studt, who gave to his performance all the wild energy he could muster. The Plague Demon realized that if it did not curb its wilful ways it might never see the Moonflower again. The Moonflower, beautiful though she was, lived off the trees of the forest flourishing as a parasite, just as we do. And I painted the pictures that were projected during the performance. Richard Harvey took my words and over five weeks created a massive and beautiful musical score. John Williams plucked its beauty like a bee collecting pollen and made honey.

Richard Gregson-Williams had implored me not to have a sad ending; not to have the audience 'gnawing on the fluted columns' as they left Exeter Cathedral, where the first performance was given. So it has a happy ending – a joyous *rejouissance*. It brought the audience to its feet and we got a standing ovation. It was subsequently played in five other venues including St Martin-in-the-Fields and St Paul's Cathedral. A BBC 2 film was made of it, too, which was filmed in Salisbury Cathedral and broadcast two years later on New Year's Eve, 1992, to an indifferent audience, probably too pissed to notice. Dammit! It is practically due for a revival.

It was powerful stuff and it still moves me! Occasionally I can hack it, and if he hadn't been so jealous of the possibility, Hunter should have written back and said: 'Bravo, Ralph!' but the swine didn't acknowledge that I had sent him a copy of the CD.

Instead, he had plans for something else.

```
24.6.91
Dear Ralph
    I think I might finally have a good idea
for us. The Answer. Filipino Bride. Dr.
Thompson Goes to Bangkok to seek Editorial
Assistant. Trial Marriage in Manila.
    Classic Gonzo. True Adventure. Desperate
search for help.
```

Thank you. Don't try to understand this thing
all at once, Ralph. I'm only about 22% into it,
myself. Right now - but usually that's enough.
    At least we could live like Caliphs
in . . .

    27.7.91
    Fuck yr. Senile simpering Ralph. You just
got a serious job. See enc. Yes. Let us rumble,
Ralph. Sharpen yr. Filthy pencil. Mahalo. Doc.

I remember absolutely nothing about that project and I
believe it remained at 22 per cent done.

It was at that time I was invited to join an environmental
team by Laura Duncan-Sandys to visit Peru, to promote the
country and tourism, and it was one of the most exciting
foreign visits I have ever made. Flying over the Nasca Lines
by helicopter, I was stunned by the majesty of it all and dubbed
it 'God's Drawing Board' and the Urubamba River 'His
Laboratory'. I wrote another oratorio making this the place
where God's plan of the universe was devised for the third
time. This time He would lay down ground rules which we
could not break. Also God was drunk and on the warpath to
keep us in line. 'Ralph's gibberish!' said Hunter, as I attempted
to get it together. He felt that I should be at his disposal at all
times, like some pocket artist, at the ready when needed. I
suppose I should have been flattered — and at least it always
made me smile!

    26.7.91
    Dear Ralph,
    I have a hideous story to tell you, and I
can only tell it once - so listen carefully.
This is it, Ralph. This is my Wisdom. This
is my Song . . . & I pass it along to You,
my smartest friend, because I love you & want
you to prosper, & also because one of these
days you will Need it.

When?

Who knows? But when it happens, Ralph — when
they come for you — (as They will. Not tomorrow,
but soon. With no warning and dumb reasons
. . . Take my word for it, Ralph. They came
for me, as you know, and Soon etc . . .)

When that happens, Ralph; When that terrible
moment comes — You must be Smart, like I was.
Yes. You will need & cry out for the Wisdom
of yr. smartest friend. As I did . . .

And you stomped on the terra, Ralph. You
will stand tall & mean in the pages of history
as Lafayette did in another time — a living,
jabbering symbol of Truth beauty and Honesty
& Wildness & Elegance & Terrible lurking
wonder. Love and Adventure ethic, all at
once . . .

That is a very powerful image, Ralph: you
standing shoulder-to-shoulder in history with
the legendary Marquis de Marie-Joseph-Paul-
Yves-Roch-Gilbert du Motier de Lafayette.

He was a warrior, Ralph. He was one of us
. . . Indeed. That is high company & very
rare air: Like diving for Black Coral at 400
feet & suddenly having to share whatever's
left of the air in yr. tank with a huge
drunken French General with a knife in his
teeth & a uniform covered with medals, no
pants and hooves instead of feet . . .

It happened to me once, Ralph . . . Keep
in mind that I am always both Ahead and Behind
you in the same moment (an eerie Truth that
we both understand in our blood and which you
have, in fact, explained more than once, in
print . . .)

So take my word for it, Ralph . . . when
they come for you, this is what you want to
say:

OWA-TYFU-LIAM Say it slowly, at first – with yr. eyes closed & yr. head thrown back & yr. throat exposed like a cheap drain-pipe – and then faster and faster in a louder and louder voice, until finally they stand back in Awe:

OWA-TYFU-LIAM

Shout it out, Ralph. Pace it like Bolero & raise yr. voice to a scream as yr. body begins to spasm . . .

They will fear you and shrink back . . . and then will you lay THIS on them: My theory of the Aborted Second Coming & How the Jews caused World War II by letting Judas get control of the Hindenburg.

12.9.92 Ralph

How could it happen, Ralph? How could it happen? How could this brainless Rat's Nest full of inbred teenage degenerates manage to Suck the very soul out of England & all it stands for?

Those dirty evil stupid little bastards! You should call a Press Conference in Kent & demand for that scum be Executed stark naked on a public Chopping Block by four hooded AxeMen to be chosen by Lottery from among the general public in all of Britain.

That would send a message to those degraded Royal mutants . . . Hell, we could hold a joint press conference, Ralph . . . maybe in the Tower of London – and call for their heads with a frenzied Mob of two or three hundred thousand boozed-up liberals, lawyers, Papists, coal miners, Artists, statesmen . . . Adulterers, Dope Fiends, Land Barons, whore-hoppers, Black Priests & tens of thousands of other decent honest Englishmen (& women) who

love the Crown & hate the horrible, never-
ending shame that these brain-dead little sluts
have heaped on it by their sleazy, filthy
behaviour that has finally gone out of Control
like a cluster of dumb Fuck-Apes with Unlimited
Charge Accounts everywhere in the World as
they wallow in filth and degrade the Crown &
mock every honourable instinct that England
ever stood for.

Shit on them, Ralph. They will suck the
Soul out of England if we don't drag them up
on the Kings Hill Chopping Block & cut off
their goddamn Heads!

I miss you, Ralph, and nothing I write
feels right without yr. foul, dehumanized ART
looming out of the words . . . And this Royal
Head-chopping gig might be exactly what we
need: We will have fun, get Rich and do the
Right thing, all at once.

OKAY. Let me know what you think. Love H

In June of 1992 he had asked me if I could create a series
of saucy pictures for his next book, *Better Than Sex.*

13.6.92
Listen, problem child - I think I can help
you out - I was doing LUST & BEAUTY before
you could get an erection.

Be in touch. LUST and BEAUTY is EASY. OK
Ralph

14.6.92
Well, Ralph . . . What can I say?
Except please convey my Condolences &
Deepest sympathies to poor Anna.

Jesus Whooping Christ, Ralph. Your hideous
Rorschach outbursts would get you locked up
forever if the Authorities ever got hold of

them — especially if you were ever accused of anything warped or wrong or twisted that had anything to do with a Woman . . .

And that could happen, Ralph. Look how close they came to getting their hooks into me — & the swine searched my home for 11 hours, trying to find something incriminating enough to take it to a Jury.

But they failed — despite a Living Plethora, on every wall & shelf, of lustfully-oriented material. But none of it was dark, Ralph. None of it bore the mark of the Marquis de Sade.

No, Ralph. I was innocent.

But God help me if I'd had this packet of brutal-sex drawings on my bed-stand. They would have hung me as a Dangerous Pervert.

Indeed. And more to the Point, Ralph — I will be shunned by My People & condemned forever in History as a Monster worse than Manson & Jack the Ripper, combined, if my 'long-awaited Sex Book' is published & spread before the Public with all my elegant, beautiful & transcendentally Romantic SEX SCENES, goddammit, are rendered utterly Foul & Sleazy by yr. aggressively wretched, Rip-Fuck, Sado-Romantic (sic) picturing (sic) of the one Lovely crypto-lustful scene that I sent you, in good faith, high Hopes & my deepest conviction that yr. ART & yr. TALENT & the mouldy remnants of yr. BRAIN will somehow Triumph over yr. Madness, yr. Foulness & the life-long Burden of yr. LOWER & DIRTIER instincts.

Or maybe it was me, Ralph. Maybe I never had to ask for anything but the Cruellest & Craziest in you & yr. heinous talent . . .

Because those were our stories, Ralph. That was the Business We Had Chosen.

And it still is, my friend — (my ancient

& honorable & highest & finest of Friends, as you know).

Yeah. The Scum will always be rising. We under stand that . . .

But so What? We will rise faster, Ralph. If they are Corks, we are Dolphins. We will dance on their heads forever — no matter how fast they rise or how powerful they feel & become — — we will always be Above them & they will always Hate us for it. Because we are the Natural Aristocrats of our Age, Ralph & there ain't too many of us — — and if they can knock us off, they win.

Ho, ho, eh? Fuck those people. They can rise, but they cannot float.

And neither can We, Ralph — if you depict me (in this, my final book) — as god's own Dirty Old Man & a dangerously Savage mauler, slasher & angry penetrator of Women, with eyes like rancid grapefruits & 14 poison dicks hanging out of my body like olde-English string-warts, & grappling with berserk women who look like Boat-hags and act like mules in Heat.

SO GIVE ME A GODDAMN PIECE OF ART THAT SHOWS THE SCENE ON THE TERRACE OF THE STOCK-MENS CLUB IN THE SAME FUCKING SPIRIT THAT I EXPOSED MYSELF TO WRITE IT. I'M SURE YOU HAVE SEEN A BEAUTIFUL WOMAN GONE MAD WITH LUST IN PUBLIC, RALPH, SO LET'S SEE IF YOU CAN DRAW IT . . . Good luck, H

Again I tried to satisfy his dismal expectations . . .

14.6.92

No, Ralph — This is not what I had in mind for my Love Scenes — it is wonderful, but wrong. Try again . . .

Fuck you, Dumbo. How about you doing that
scene that I sent you (the Stockman's Club) –
    Yeah, Ralph – maybe you can't illustrate
Lust & Beauty at the same time. Maybe you're
just a dirty old man, eh?
    OK – you have 24 hours. HST.

Hunter hated me to draw images of any kind for others,
particularly the cover for the unauthorized biography, *The
Strange and Terrible Saga of Hunter S. Thompson* by Paul
Perry, the man who brought us together to cover the Honolulu
Marathon in 1980. I couldn't decide if it was humour, jeal-
ousy or plain insecurity.

    10.7.92
    Dear Ralph,
    You have caused me a lot of trouble with
Various People by doing some kind of Cover
Art for either Peter Whitmer's (or Paul
Perry's) BOOK on ME, which may or may not be
scheduled for Publication soon.
    Shit. How can I know. Everybody involved
has lied to me about these cheap, Soon-to-be-

Buried gossip books about random strangers'
comments on My Life — all of them with an
utterly different Bias & a different set of
quotes & characters — & I have already
contracted with the NY Times Book Review to
wait until they're all done & then review them
as a whole, for whatever kind of half-bright
scum they will seem to be at the time(s) of
their various publication(s).

I'm sure you had yr. own deeply personal
reasons for endorsing (at least) One of these
sleazy Paparazzi (sp?) outbursts . . . or
maybe Two (2) of them or maybe all Three (3).

Who knows, Ralph? You were Smart, once —
but then you got into Wine. And, after that,
'things changed' — as they say in the old-
time romance novels . . .

But who could blame you, Ralph? You faced
unnatural Pressures. You had a wife and child
& a Castle in Kent with a Swimming Pool that
needed constant attention to the Algae-Filter
— and then, of course, there was always yr.
holy Need for Money.

God knows Why, Ralph. I have, myself, dealt
with what people like You have called (in
print) 'the Intolerable Pressures of Being a
Life-Long Dope-Addict,' & I know exactly what
you're going through.

It is hideous. And I want you to know that
I have Forgiven you in Advance for whatever
dumb gibberish you might have uttered, for
money, when these rotten little Paparazzi were
putting the squeeze on you.

Jesus! Ralph ! ! Jesus . . . I know you're
Weak — but so What? My only hope is that you
have said & drawn & gibbered whatever is Right
for you, in yr. final wisdom as a Man & a
father & a vaguely-diluted Artist.

Yes. You must Save Yourself, Ralph. Nothing
else matters . . . Don't babble like a goddamn
stupid brook. Why not? We were, after all,
professionals.

So long. Hunter.

I responded to show my disquiet. Our faxes were often
vicious, but were spiced with hilarity.

11th July 92
Oh, Hunter.
Some of your biographers have been pestering
me. I have been accused of being mean, obstruc-
tive, snooty, an enemy to American Literature,
a geek with cheap ambitions of my own, and
worse, a PERVERT. I have even been pressured
into letting Paul PERRY use one of my drawings
on the cover of his book — "because I owe him!"
Your biographers want me to verify why you
were really in Kentucky. That you kept Juan
in a cage with a drug-crazed minor [sic] bird.
Conducted vivisection experiments on stray
dogs. Peacock ABUSE.
The 'Black Boys' episode in Kinshasa. They
are offering good money to tell all. LAILA
has also been got at and she knows plenty.
How much are you going to pay me NOT to
tell??
My children are starving & I have RICKETS.
I am your friend — Ralph.

But it continued:

14.7.92
My dear Hunter
Your most terrifying FAXES, as always, come
like a 'knock on the door in the middle of
the night.'

I tremble as the fucking white paper begins
to shudder like a death rattle out of my
gunmetal grey machine.

'What have I done wrong now?' dribbles through
the gummy lower part of my jaw. My teeth lie
in a jar upstairs, four of them at least. The
rest can barely keep my mouth in shape to utter
anything coherent. All I know is the fear inside
of your constant accusations.

NOT BECAUSE I AM GUILTY OF ANY SPECIFIC
CRIME EITHER — only because I am a friend who
is here to be abused — particularly in the
MIDDLE OF THE NIGHT by YOU who seems to take
a particularly PERVERSE delight in feeling
WRONGED by your wretched friends, GOD help
them!

Here, for the record and for whoever gives
a shit is what I have done — RIGHT or WRONG.

I have allowed Paul Perry to use one drawing
on the cover of his forthcoming biography of
you. He wanted more, much MORE and I refused.
If you like, it was a gesture to a one time
colleague — maybe a foolish one, but a gesture
nevertheless. According to your FAX I have
obviously collaborated with the 'ENEMY'!

I have had many such requests for all sorts
of shit from your army of 'biographers' and
I have refused every one.

ANNA said it was wrong to allow even one
drawing to be used and as always ANNA was
right. But I did not listen — I was trying
to be nice. Yes, nice. And all I do is end
up being pathetic.

I was going to come and see you in August
— if I could get MTV to pay my fare, so that
they may film us together chewing over the
fat like old buddies, but perhaps that too is
a violation of your privacy. I am at a loss

to know what is right - but I would like to
see you.

Advise me, if you must, but don't reject
me. You know you have been more than instru-
mental in my destiny as an artist and even
as a human being and that I estimate you
above all living writers. There. There's a
confession - though I'm not so sure about the
human bit.

I was going to keep this short but I have
wallowed a moment inside the written word like
the buggering fool I am in the quiet of the
night.

I probably won't sleep now waiting for that
knock.

What are we going to do with you? You probably need friends like me, and particularly
ANNA who perceives the consequence far beyond
the average. ANNA is always right.

If you want to reply, pick up the phone
and speak. FAXES are FUN but they are brutal.

Cazart as you used to say. Love Ralph.

Simon Kelner, who is now the editor of *The Independent*,
revived the idea, this time for *The Observer*, of getting Hunter
over to cover the royal occasions of the Highland Games and
the Queen's holiday at Balmoral or anything that would put
Gonzo in the front line for a story about royalty. Whatever it
was, it would be a hoot, even if it went wrong because that is
where Hunter operates best. He wrote to me in a more jovial
mood . . .

31.8.92
Ralph
Don't worry. I'm on my way. I'll be in
London on Thursday & we will crush these pigs
who dare to censor & defame you.

I will be at The Metropole Hotel, writing

a piece on the Royal Family for the Observer.
   But I will need the White Death, Ralph -
& I'm sure you can help me.
   Right. Please advise quickly. Today. Time
is short. Okay. Thanx. H.

The *Observer* 'Royal Visit' story was covered by my good
friend, journalist and writer Robert Chalmers. He has allowed
me to draw liberally from the notes he took as events unfolded.
Since 1996 he has been with me in Aspen on several trips to
interview Hunter. He became a real friend of Hunter's who
estimated his abilities and intelligence above many who had
entered the kitchen at Owl Farm and attempted to plumb the
depths of the darkest soul on the planet who would not be
plumbed. Robert's northern spirit broke through. He disarmed
the bastard long enough to denude him of the information he
required while the barriers were down.

4.9.92: ARRIVALS HALL, GATWICK AIRPORT

'Hunter is one of the first passengers to emerge from the
arrivals hall having negotiated the green lane with a brisk-
ness that says little for customs formalities at Gatwick. An
ungainly man, he wears grey terylene trousers, white training
shoes and a baseball cap decorated with the words "Polo Club
of America". At his side, Nicole, a small, blonde woman in
her mid-twenties, struggles to control a trolley which groans
under the weight of six large cases, half of the doctor's
equipage for the five-day visit.

   Dr Thompson, whose breath suggests that he has not
stinted himself on the in-flight service, marvels at his miracu-
lous delivery from the hands of Customs and Excise, then
begins to lament the fact that he has not brought more drugs.
"It is the curse of cowardice," he mutters, blocking the esca-
lator and opening a small bottle of whisky. "Cowardice, terror
and guilt."

   As they make their way out of the concourse, Nicole
boasts that she asked Hunter to pay her two hundred dollars
for every airport she had to take him through. She recalls

problems at Saint Louis, where he attempted to hide in the gents with a view to bolting for Aspen: "I lost faith in the airplane," Hunter explains.'

At the time Robert, only recently acquainted with Hunter's travelling habits, took this to be a joke. And Hunter's description of his TWA luxury airliner as a 'cattle truck' did little to assuage his doubts about their visitor's accommodation. The original plan to book him into Brown's in Mayfair had to be abandoned when it emerged that, on his last visit, Hunter had overturned his bath, bringing down the ceiling of the room below. Instead he was to stay at the London Metropole, a large hotel near Marble Arch where he had been booked into the best suite at three hundred pounds a night.

Pushing the trolley with its substantial load of what Robert later discovered to contain mainly video equipment, hashish and weapons, Hunter headed for the exit ramp. On the brow of the incline he stopped, restless because he had finished his whisky, and fumbled for his cigarette packet, taking one out with his right hand. He searched for his lighter with his left, leaving the trolley to freewheel on its own. It gained momentum and headed straight for a plate glass window ten feet above street level which it somehow failed to shatter. The only sign of contrition from Hunter was a slight wobbling movement that might have been interpreted as a half-hearted attempt at pursuit. 'These things,' said Hunter, pointing at the trolley, 'tend to wander a bit.'

Half an hour later they were travelling by cab through the rush hour traffic, Hunter in the back, wheezing and peering through clouds of marijuana, occasionally yelling remarks at motorists and commentating on the views through the window: 'Poor fucking dingbats, slobbering idiots roaming in the streets; doom, death and decay.' Passing a man sitting on a stool by some traffic lights wearing nothing but his underpants, Hunter shouted: 'Death to the weird.' Fretful because he couldn't find his spare bottle of whisky he issued urgent requests for other things he couldn't find, like beer and cocaine: 'I'm smoking all this grass and I've got nothing to brace it,' he complained.

At Purley Oaks, Hunter made a surprising announcement

that he would 'like to buy a house in Purley', admiring the semi-detached houses on the main road which, he said, 'have class. Check out the real estate offices,' he told Nicole. 'Place in the country,' he muttered, 'place in the town.'

As they approached Waterloo, Hunter performed his first and, as it turned out, his last journalistic activity of the trip when he spotted a dilapidated shop surrounded by listless old age pensioners. He noted down the name on the door: Waterloo Action Centre.

Hunter had become increasingly ugly about his need for alcohol, so they made a wide diversion to the Fox and Anchor pub, next to Smithfield meat market, where the opening hours catered for the meat porters who worked there. The barmaid enquired whether Hunter was famous since he acted, she said, as if he were. He ordered two pints of orange juice, two large Bloody Marys, a pint of bitter, coffee, a triple Scotch with ice and a vodka and tonic. He ordered an English breakfast: 'Bring every condiment in the house.' He smothered his plate in an uneatable layer of mustard, pepper and salt and left it largely untouched. As Robert noted: 'Breakfast with Hunter is descending into amphetamine psychosis.' In the gents, Hunter retreated to a cubicle leaving the door wide open and snorted cocaine vigorously. Over breakfast he rambled about not understanding his brief, not being taken care of and not knowing what he was doing there. He emerged, snarling, from the pub, where you could eat well for five pounds, yet the bill came to eighty-nine pounds.

Back in the cab Hunter was adamant that he would not go to Scotland. He had no brief. He couldn't work with 'fucking English dingbats'. When Nicole attempted to reassure him, he told her that she didn't know 'what the fuck' she was talking about. He stared glumly at his shoes, made no eye contact and took large slugs from his whisky.

With much relief on Robert's part, they arrived at the Metropole Hotel. Nicole exited the taxi 'like a bullet from a gun' while Hunter's approach to check-in formalities took the form of slumping down on the marble floor against a pillar and muttering to himself. Robert walked over to ask him what the problem was. Hunter replied: 'You guys are taking me for a fuckin' asshole,' though his spirits seemed to revive at the sight of the large bar in room 828, the Stevenson Suite, and he immediately began to search his luggage for, not medication or underwear, but a large novelty hammer. When lightly struck, it emitted the sound of broken glass. Lots of practice with the hammer and the intake of whisky, vodka and Grolsch seemed to restore him. He opened the French windows, climbed over the safety rail and teetered on the balcony, peering out over the Edgware Road. Nicole complained that she hadn't been to bed since Sunday and said they wanted to be left alone for an hour. Robert left them.

Hours later, Robert and Simon were still waiting in the bar downstairs. By then Hunter should have been in a taxi to Heathrow to catch the afternoon flight to Aberdeen. Nicole appeared and seemed to have recovered from the events of the taxi journey when, as Robert noted, she had been in the condition known to shipping forecasters as 'precipitation within sight'. Hunter had issued a list of requirements that included press clippings, a manual typewriter and someone to meet him in Scotland. There had also been mutterings about 'dingbats in Canary Wharf', the need for a pair of binoculars and, surprisingly, interest in the Post Office Tower. Nicole was 'relentlessly plausible' and confirmed that they would take the 7.55 flight and also that I would meet them off the plane. I could have done this since, by a strange coincidence, I was working on a series of silkscreen prints and etchings at a print workshop in Aberdeen. Meanwhile, the receptionist had said that they had had 'one or two slight problems with Dr Thompson'.

Robert admits that their next big mistake was to believe what Nicole had said. At midnight they called it a day and left. There was nothing they could do. The door of room 828, which the staff had begun to pronounce in the tones of a biblical 666, was firmly locked.

I called Robert in the morning to say that Hunter had phoned late the night before. He was still in the Metropole, said that he was unwell and that the suite was 'halfway between a fucking Bauhaus and a dog pound', and that he had festooned his door with 'Do Not Disturb' notices. Robert, having returned to the Metropole, received no response from repeated knocking on the door and the room telephone had been disconnected. The security man, Dave, said that the occupant of room 828 had initiated several lively telephone discussions with reception, who seemed as eager as he was to see Dr Thompson board the Aberdeen flight. He hadn't left his room in twenty-four hours and the hotel computer print-outs revealed that he had ordered three Bloody Marys and had made three phone calls, all to London numbers. These turned out to be British Airways, United Airlines and TWA.

As Robert's and Simon's vigil at the hotel bar became more protracted, they developed a healthy rapport with the staff, who offered several services not listed in the hotel brochure, including breaking in and impersonating a chambermaid and giving him 'a good bollocking'. The last Aberdeen flight had departed and Hunter was reportedly out cold. I got a phone call from an anxious Simon asking me how they could get Hunter out of his room. 'Room service,' I replied, 'send up a trolley-load of stuff with a note saying "Guns, whisky and the Royals await in Scotland," signed Mr Skinner,' (a shady character who appears in *The Curse of Lono*, looking for drugs – '"You mean drugs?" Ralph said finally. "OF COURSE I MEAN DRUGS!" Skinner screamed. "You think I came here to talk about art?"') 'knock, and when he says, "Who's there?" say "Room Service." He will open the door and you're in.' In the event, Hunter opened the door, yanked the trolley in and slammed the door.

The following morning there was still no response to their frantic knocking. By now, they were seriously concerned since Hunter had complained to me by phone that he was weak from vomiting. He could be unconscious or dead. The sense of alarm was augmented by the thought that room 828 might be littered with enough 'White Death' to 'give it an appearance of Anchorage at Christmas'.

Late that night, a chambermaid was sent in and reported that the room was in considerable disarray but empty of clothes and bags. On the sofa, all Hunter had left was an airline comfort blanket and a Metropole Room Service Menu inscribed in his own hand. 'This is the nearest to written reportage the doctor is to come in the whole trip,' Robert wrote, 'scrawled on the cover, in pencil, was the single word "Dorthe".' The valet recalled sending up a porter at seven-thirty that morning and the doorman remembered an 'unusual man' asking for a cab to Gatwick.

Disconsolate, Robert and Simon met Simon's wife and three-year-old daughter at a restaurant near the Savoy called Smollensky's that caters specifically for children. The atmosphere of bonhomie was not quite in keeping with their mood but gradually they started to revive. They had bottles of champagne and took turns with the comfort blanket. Every song they played in the restaurant seemed to have a particular resonance with their recent difficulties. These included 'Home on the Range', 'Take Me Back to the Black Hills of Dakota' and 'I Thought I Thaw a Puddy Cat.'

Shortly afterwards the *Observer* held a party for Hunter at the Metropole. In the absence of the doctor, the guest of honour was P. J. O'Rourke, who explained 'with uncharacteristic coyness' that he once spent a night on 'drink and other things' with Hunter and he was in bed for a week afterwards. Robert ends his notes with a few moments of reflection:

> It was one of several moments that weekend when I had the feeling that Thompson was somehow capable of creating an atmosphere of infectious strangeness that persisted after he had left, as teenage girls are supposed to act as catalysts for the activity of a poltergeist. For the next few days he was sending mad faxes from Aspen, castigating 'that evil treacherous dingbat Robert' and 'the whole bunch of neurotic cultural elitists and spiritual Nazis who control not only *The Observer* but the whole goddamn British press, and the whores and the hoodlums, paid-off fascist sluts from Jaguar and Rolls-Royce and helpless BSA, who worship and fear and wallow at the feet of your bogus, hare-brained royal family.'

Friends say that abuse is Hunter S. Thompson's

way of telling you he likes you, a belief we clutched to reading his final fax: 'Do not delete or cut any of my copy. I'm tired of your shit-eating censorship. Fuck you. HST.'

Looking back on this weekend, I feel like a member of a struggling first division side which has just lost 6—o at Anfield: desolate, routed and in shock, but nevertheless convinced that, had we to repeat the experience, we would not make the same mistakes again. It was especially frustrating to have come so close to getting him up to Scotland; we consoled ourselves with the knowledge that at least one previous attempt to bring him over had failed to get him beyond the departure lounge at JFK. Perhaps it was all for the best.

'When I heard he was in London,' Ralph Steadman had said, 'my blood ran cold.'

Hunter wrote to me on his return to Owl Farm.

Ralph
You sleazy shit-eating whore! You are the leader of the Jackal-pack in England, now. Simon Kelner is the lowest & dumbest kind of scumbag editor I've ever worked with. He has even faked a photo of me with the Queen on pg. 1 of the horrible sleazy attack on me that The Observer calls 'an article by HST.'

The shithead re-wrote most of my work/words & cheapened my whole literary reputation in the UK & all over the world for the rest of my life.

Not even the sleaziest & cheapest editor has ever rewritten my words & my lines & changed the meaning of what I wrote.

Jesus! The British press has always acted like scum - but the tabloids at least have an excuse which I understand & even respect

in some odd professional way — But The Observer
does not & neither do you.

Ah . . . but so what, eh? Win some, lose
some.

But I was right about the certain degrad-
ation of any trip to England & thank Christ
I didn't come to Kent.

You are a treacherous back-stabbing commer-
cial artist who will do or say anything to
make a cheap pound.

You couldn't get work in Germany or even
Italy — & as sure as hell not over here. Jann
was right about you, & I will not fight that
battle again.

The 'article' in The Observer is less than
33% as I wrote & submitted it. Even the photos
are faked.

Kelner is a liar and a fraud who will get
The Observer in deep trouble by trying to
lower it to the level of the Mail & the
Express & the low-rent National Star. He won't
last long.

A year later, proposals for mad projects still came up:

27.12.93
Dear Ralph,

Thank you for the wonderful convration TALK
Talk yes, this is my new COMPUTor Computeeer
. . . yes. youknow what I mean.

Okay. I KNOW IN MY BONES that this VISIT
TO RUSSIA: What does it all mean? FEAR &
LOATHING By Hunter S. Thompson and Ralph
Steadman . . . CONVERSATIONS FROM THE STRANGEST
PLACE IN THE WORLD . . . REPORT FROM THE
SHATTERED SOVIET EMPIRE by The Most Famous
Foreign Correspondents of the Twentieth Century

Don't worry. This is definitely our kind

of story. We will beat it like a gong. I know
many cruel people in Russia. We will join our
DRIVER in Berlin in late April, and then to
Moscow. We will film it. Okay. Let us rumble.
    OKAY,
    HST

He had some fun at my expense on hearing a story. But I
was vindicated!

    3.1.94
    Owl Farm
    Dear Ralph,
    Some giggling yoyo artist (sic) from the
New Yorker was here for a few days over Xmas,
and when I told him he should get out of town
& turn his job over to "a real artist like
Ralph Steadman" he laughed out loud & said
you were so real that the New Yorker made you
re-draw a portrait of Barry Diller four (4)
times until Diller finally approved it. I
called him a lying swine but he said the
horrible proof was in "last week's New Yorker."
    "They made him do it over and over", he
said, "until Barry liked what he saw."
    Jesus, Ralph! What can I say? It was a
horrible thing to hear, and the way that he
chuckled, which is true. But it was also
"the nineties", as I recall, when you did
that hideous 8-ball portrait of U.S. supreme
court justice Clarence 'Long Dong' Thomas.
It was the ugliest thing I'd ever seen you
draw and it made Jann grind his teeth until
the enamel cracked off - - But neither one
of us even thought about CENSORING it, or
DROPPING it, or sending it to THE JUDGE for
his Approval.
    It was your Art, Ralph, and that's how it

ran in RS, for good or ill. Buy the ticket, take the ride. I would have Killed anybody who said it was too mean-dumb & ugly to put in print - - which it was, of course, but So What? I would have gouged out my (?) before asking you to "draw it again" - - maybe with higher cheekbones or a slightly more flattering skin-tone. And Jann took a shit rain of outrage about it - - not only from the Negro community & the Hip-hop people, but also from rich Japs in the music business, and even from Jane, who hated it.

SO I'M SURE YOU UNDERSTAND, RALPH, WHY THIS WRETCHED GOSSIP ABOUT YOU AND BARRY DILLER CAME AS A SHOCK TO ME, AND WHY I CALLED IT A PACK OF LIES. I knew it was not true - that you would never submit to vainglorious censorship by the rich. It would mock everything we've always stood for.

"Fuck off," I told him. "You're not fit to shine . . ."

3.1.94

Hunter,

Hold off on chastizing Steadman. IT was Gerald Scarfe who redid Barry Diller 4 times. MY mistake. Never mind.

Thanks for the words and pictures. May be in the Jan. 24 issue, if I meet the deadline.

Yrs.

Bill Greath.

3.1.94

Ralph!

Hot damn

This swine has admitted guilt.

It was Scarfe. Not you. Why do they lie about us?

Okay — now we can go to Russia without
shame.

Nobody censors your art, Ralph.

Not even in Moscow. I will cripple them!!
OK Later. H.

3.1.94

Well, HUNTER — I'm a little surprised. You
know better than to listen to gossip.

There is only one answer to that kind of
GIGOLOSHIT — who the devil is Barry DILLER?
And I have never worked for the NEW YORKER
in my life — because I'm just not good enough.
Worse — not even prepared to stand in line
and submit my works — for fear of rejection.
It is a terrible failing of mine. I suffered
too much Playground bullying to subject myself
to pain of any kind in later life — even
though later life is upon me — us, now.

Just one thing, Hunter, I won't draw
Politicians now — haven't done for 5 years.
I reckoned I wouldn't invite any of them to
dinner, so why the fuck should I draw them?
Particularly with my PROBLEM — they NEVER
bought me a drink — and now I come to think
of it — neither did you. But what the Hell!
You always never bought me one with style.

Fuck off, you drunken geek! — you always
said. Buy your own goddamn drink!! And it
always made me want one even more. But that
was the 70s, and now it's the 90s, which is
only the 60s upside down, eh! We can still
trail blaze like Victorian Explorers with
Dysentery — so it's a NEW RUSSIAN SPRING.

A chauffeur and four interpreters — BALKAN
philosophy.

How d'you like, WHISKY BAROQUE — An Illicit
Romance — for my whisky book title???

Anyway, that's my problem and I have a
hundred others - problems and titles.

You still look frighteningly SANE in your
latest photos. Is it a sign of weakness, I
ask myself?

OK. You've had it for today.

Love Ralph x

3.1.94
Thank God!

Ralph! I knew the bastard was lying about
you & B. Diller. But it is evil hateful gossip,
none the less - & I will do my best to squash
it. Okay. We're on for Russia.

H

But, times were moving on. In June 1994, Hunter's old
nemesis, Richard Nixon, died. Hunter marked the occasion
with a note to me:

D. Day Death to the Hun

Dear Ralph,

It has taken me a long time, old friend,
but I think I have finally made you proud of
me. Mr Nixon, he dead . . . . . . others
wavered and waffled. But not me, Ralph.

I beat him like a re-headed mule, & doing
it made me happy. No winery or foppish eatery
paid me extravagantly to do it - but I did
it anyway - just for expenses & the pure love
of flogging that perfectly monstrous bastard.
He brought out the best in us, Ralph, all the
way to his wretched illegal end - and I will
miss him . . . Let Hell receive an honored
guest: Richard Milhous Nixon is finally laid
to rest. He was wrong.

Maybe I was queer for the brute, but so
what? He kept me angry & alert for 30 years,

and he was always there when I needed him
. . . But he is gone, now, and soon I will
follow. My life is meaningless without
him.

And I shit on yr. chest, Ralph, for leaving
me to battle these demons alone. If you get
what I'm saying . . .

Maybe you could dash off some art to go
with these (enc.) 4 pages of my final judg-
ment on Clinton. Read it & weep. Rumors of
my death are vastly exaggerated. Love, Hunter.

10.7.94
Dear Ralph,

Yr. theory of IRREGULAR IMPETUS ranks with
Krapp's Last Tape as perhaps the finest
outburst of Human Understanding in the 20th
Century - - yr. ART, of course, is like being
stabbed in the ass from behind in a crowded
pub.

But I am more concerned now with yr. travel
plans - - especially LOUISVILLE & BOULDER,
which should put you right in my fiery greed-
crazed neighborhood.

You must visit, Ralph. This may be our
Last Chance.

Let me know ASAP. Thanx. Hunter.

I was still working for Oddbins, trying to pay back the rent
we had stolen to get us through the eighties. I was assigned
to see the whiskey distilleries of Kentucky and Tennessee. We
arrived in Louisville, Kentucky, on 12 July and decompressed
at the Brown Hotel, named after its proud founder, James
Graham Brown, who had made his millions in timber and the
horse trade.

Hunter must have known of Mr Brown, who had died in
1969 leaving a foundation to foster the respectable face of
Kentucky, its image of well-being, quality of life and all the

decadence that Hunter had been at pains to unmask when he met me at the Derby. The hotel was elegant, writhing with carved ceilings, and dripping with the opulence of lush velvet curtains. It was like coming home – a full circle from my first meeting with Hunter in 1970. We hadn't stayed at this superior hotel then, however, working as we were on a shoestring for *Scanlan's*, whose owners treated us as foot-soldiers with an insurgent's grudge. We may as well have shacked up with the horses in the stables. God knows, horses are treated to an exalted life of luxury and care, like exclusive citizens that reflect the priorities that only a Derby town can show. It felt weird. People seemed nice enough, helpful and hospitable. I could not detect anything that resembled the sick vision that rankled in Hunter's brain. He was only trying to get his own back for the abuse and rejection he claimed he had suffered years earlier for being such a goddamn awkward cuss himself. The swine could have been lying and I fell for his story of bitter revenge. He had probably been thrown out of every decent establishment in town after leaving untold damage and drunken mayhem in his wake, thieving as he went about his twisted purpose. Hmmm! It only required a total stranger from half-way around the world to believe him and also, remember this, he recognized a fellow rejectee who also lied and swore about his treatment at the hands of those who thought that they were better than him – and worse, that he himself had exaggerated the foul behaviour of the older establishment of his home town in North Wales. All were corrupt and our common destinies were forged from that moment on – and our crusade!

So, there I was, back in the same backyard for malcontents, looking for trouble. BUT THIS TIME I WAS GETTING PAID FOR IT!

The whiskey trip was a highlight and even Hunter approved when I told him. It touched his pride, in some way. We visited Jim Beam's Distillery in Bardstown, home of Stephen Foster, who wrote 'My Ole Kentucky Home'. The Bardstown Oscar Getz Museum of Whiskey History stood proudly in the middle of town, manned by weathered ladies

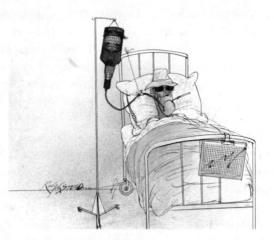

who seemed proud of what they were minding but keener to cook with the stuff than drink it.

Then it was Maker's Mark, the Distillery in Loretto, a beautifully restored nineteenth century building inside which they make a Southern version of Scotch known as Bourbon Old Style Sourmash. Still no one offered us a drink. On our third day we went to have lunch at Miss Mary Bobo's Boarding House in Lynchburg, Tennessee, where you most definitely can't get a drink.

The following day we were taken by the nose to Tullahoma, home of George Dickel's Distillery. I've heard tell that, with one hand tied behind his back and one foot in Cascade Hollow, ole George could make the finest, smoothest, sippin' whiskey in the world and you can drink it anywhere in the world but in Tullahoma. All we got at the distillery was a lousy cup of coffee and a piece of whiskey-laced draught excluder called Dickel Cake! It was hideous and we couldn't get away soon enough.

But forget all that. You can read about that in my book on the stuff − *Still Life with Bottle*. We headed for Nashville and a proper drink in a bar with a live musician at your service. Travelling south towards Nashville we passed a billboard sponsored by the Department of Health. It read: 'VIRGIN − TEACH YOUR KIDS IT'S NOT A DIRTY WORD.'

We had breakfast in the Cowboy Star Café and gawped at a karaoke country singer whose only other audience were waxwork figures of familiar stars with fixed expressions, unable to leave. There was a stuffed horse seated at the bar and Gordon Kerr, who had accompanied us on this trip for Oddbins, sat side by side with the horse on the next stool, talking to it, for Christ's sake!

On the Sunday of that week, five days into our search for whiskey on Hunter's home turf, we took a walk to a Cracker Barrel Coffee Shop beside a wide road, not exactly a freeway. As we dashed to cross over between traffic I saw a woman's shoe on the tarmac and decided to rescue it. It looked ominous and, being a conscientious innocent abroad, I looked around for the second shoe. I found nothing, nor did I find a leg or even a foot. Was there a missing persons bureau in the town, I wondered, and if so, what about her head? Was she beautiful, lost or alone? Did she also have a body or maybe only one tit? Or is there a woman with only one shoe staggering around Nashville blamelessly asking passers-by about a shoe? She could be in danger! These questions came to mind as we ate our mountains of bacon and eggs over-easy, we watched women, with arses like unhealthy elephants and hips like bumper cars, waddling over to the pay-out desks and grunting their approval at the recently devoured stacks of honey-dripped pancakes through asthmatic camels' wheezes. They kept coming, herds of them, barely able to lift up their equally stuffed handbags to get out their purses. This was where Hunter learned his filthy eating habits, I thought.

Gordon flew back to England while Anna and I went on to our next destination – Lexington. We thought that since we were in the neighbourhood we would call on Joe Petro III, a silkscreen printer, who had been writing to me for some years. Joe was about to go on a weekend with his girlfriend's children but he sounded delighted, and relieved, when he answered our call. We would have missed him if we had called ten minutes later. He came to the airport to pick us up, having cancelled all for us. Apparently we saved his life and changed the course of history!

Joe's small house is full of dogs and a parrot called Bub. It has a similar squawking pattern of speech to Hunter's mynah bird, Edward, and calls out Joe's name throughout the day, as well as delivering various philosophic outbursts about the American Constitution.

Joe makes silkscreen prints. Including ecology images for Greenpeace. We didn't mess around and planned to make prints of a couple of my images down in his basement. *Bats over Barstow* from *Fear and Loathing in Las Vegas*, seemed like a good start and by lunchtime I had completed the black overlay of the crazy maniacs driving across the desert towards Las Vegas to wreak havoc. By the end of the afternoon I had made two transparent, acetate overlays of background desert colours and had added the finishing touches. I was looking forward to printing it and next day we were off to his downtown studio workshop to get going. Anna went off to visit Shakertown with Joe's girlfriend, whom he called Polky, though her real name was Jessamine. She spoke with a heavy Southern accent which sounded delightful.

Joe is a thorough and conscientious artist and his workplace is always immaculate. Layered racks stood empty, awaiting fresh editions of prints like open-mouthed crocodiles. The colour layers were laid down first using a rubber squeegee through a photographic stencil ironed onto stretched silk and held tight inside a large rectangular wooden frame. With a single stroke Joe would transfer a gloop of poured ink through the silk and onto the paper which is distributed according to the stencil. The ink, transferred by this most direct of printing methods, was then slipped into the stands which kept each sheet separate to allow for ink drying.

During the last hundred years this technique has been used best in the making of advertising posters and political pamphlets declaring everything from Marie Lloyd at the Hackney Empire music hall in East London to 'Down with Everything' and 'STOP THE WAR!' It is simple and direct – the common man's printing process. We printed seventy-seven copies of *Bats over Barstow* for our first edition and decided to issue a signed edition of sixty-seven. That was that. An

idea was born and on subsequent visits we collaborated on many such prints, the most ambitious being the huge *Lizard Lounge* print from *Fear and Loathing*, measuring 38x50 inches, in a signed edition of seventy-seven. That one was a ball-breaker and there was ink everywhere. We have been working on these prints for more than a decade and I see no reason to stop now.

One can derive the same fun from print-making as from making mud pies and great subtlety can be achieved through the use of transparent inks, half-tone screens and even accidental colour combinations, which is often where the art hides. In fact, the art of stencilling goes back to the Welsh caveman and the Australian aborigines who used to blow white and brown ochre across their hands onto a cave wall, to spruce the place up a bit. Sadly the computer is taking over. Unless you get ink on your pants, you never get involved. Prints are just not prints anymore and it is difficult to know what can be called a hand-made image. No blood, sweat and tears, so no tactile surfaces which are so beautiful to see and to feel.

From Lexington, we moved on to Seattle to visit a company called Starwave, an enterprise owned by Bill Gates's early partner, Paul Allen, who had taken it upon himself to transform my book *I, Leonardo* into a CD-ROM, in which you would be able to explore Leonardo's studio and interact with the minutiae of his intricate and ingenious life. It was never finished and technology marched on to new fields of business endeavour. At least my mural of Leonardo da Vinci still exists on the wall of their offices, which were subsequently taken over by Expedia. They had the wall insured.

Next stop Aspen, to see Hunter again and attend the wedding of his son, Juan.

Hunter was waiting at the airport respectfully this time, as Anna was with me and he knew from the seventies onwards that she was aware of his bellicose ways and, like a stern schoolmarm from Decatur, had mentally put him in his place with her A-level English charm. He was beginning to look a little older, sweatier and fuller in the face and his walk was becoming more of the caricature of what I remembered when we first met in Kentucky back in 1970. But he was always at his most apologetic and hospitable at times like that and his welcome was warm and huge, in a bear-like way. He had brought along his stuffed wolverine which sat with a perpetual grimace on its face in the back of the huge red Chevy.

We loaded up our baggage and Anna sat with the wolverine

on her lap, a box of beers on the seat next to her. Hunter slammed the trunk with a flourish and then started looking for his keys. He realized he had locked them inside the trunk. We agreed to mind the car while he got a cab back to Owl Farm to get another set of keys and we stayed behind to guard the red shark and the wolverine. Several people quizzed us about the situation – just visiting? But, more specifically, they asked what year the car was, what was the horsepower . . . but not a word about the wolverine.

It didn't take Hunter long before he was back with a spare set of keys, but none for the trunk. He promised to take us to meet the taxidermist who had prepared the wolverine for him. It was to be a special treat. But first he swerved us up to Woody Creek and the Tavern where we stopped to have a couple of drinks. It was so nice to be back and, thankfully, nothing had changed. Except for an art gallery run by a man called David Florian who insisted on opening up to show us some of Hunter's artwork. At last he had bitten the bullet and wanted to show me how art was really done, 'but remember Ralph,' he warned, 'it's not art until it's SOLD!'

He had taken images of Mickey Mouse, Richard Nixon and himself with shaved head from his Sheriff's Campaign from the early seventies and blasted them with a 12-bore shotgun before framing them behind glass. I was very impressed, but I reminded him that William Burroughs did his 'shot-art' first.

Hunter was sharing the gallery with Joe Andoe, an artist who had produced some formidable life-size horse paintings which impressed me a lot and reminded me of why we had made this trip: *Polo is My Life* – an article for *Rolling Stone* that we talked about turning into a book – as well as Juan and Jennifer's wedding.

Hunter wrote to me regarding *Polo is My Life*, (he had also written a wonderful foreword to my book *Gonzo the Art*):

```
Okay, Ralph, the Joke's over. This time
you've gone too far. Your brazen slime-coated
efforts to cheat me out of Fifty Percent (50%)
```

of earnings on this Book is a horrible joke,
and my Lawyer will take it personally.

It might be easier to just settle this
thing with a one-time payment (to me) of 50
thousand Pounds. That would be easier than
haggling about royalties later on, eh?

Anyway, here's yr. stinking Foreword. Fuck
with the Bull, get the Horn. Why not? This
could be good Publicity. The Brits love scan-
dals. Let's whoop it up.

Yr. bleeding Host,

Hunter.

We were off to Owl Farm, taking the 180-degree turn up
Woody Creek Road by accelerating and urging the lurching car
up a grass slope between the two roads, Hunter's short cut. He
always managed to do it before grinding and lurching along
the couple of miles up to the farm, which I thought looked
tidier and more organized than on our previous visit. There
were sprinklers watering the lawns and the peacocks were still
in residence. The gateway was adorned with a pair of rusting,
iron, origami vultures and logs for the winter were carefully

stacked along the front fence. Generally, the place looked as if it had had a make-over and felt good. Deborah Fuller, Hunter's faithful assistant of many years, welcomed us warmly. It was good to see her. Through the years she had kept him focused on the essentials without interfering with the natural flow of what obsessed him and that is a precarious tightrope. We looked at the view. Owl Farm is situated opposite a fabulous view of an Aspen tree-lined ridge behind which the higher mountains of the Rockies peep and majestically define the sky's edge.

Anna rarely sat in the kitchen, except in winter. Instead, she preferred to lounge outside in the sunshine on the verandah with the peacocks and humming birds, drawing the view and writing and sketching in the diaries which I often peruse for long-forgotten details. Eventually, we managed to open the trunk and transfer our baggage to another gasping whale of a coupé which we used during our stay.

That night we had supper with Ed Bradley, the journalist and broadcaster, and the collection of friends who make up Hunter's Aspen crowd – Cilla Hyams, Quentin and Gayle from whom we were renting a cabin called the Carriage House in Aspen, Gerry Goldstein, his personal lawyer (Hunter always had an entourage of them and I think he was secretly observing them and their personal habits), and his wife Chris and a couple of old friends of Hunter's called Semmis and Dibble. The supper went on well past midnight but Hunter was still determined to take us to meet his taxidermist, who lived in a white box of a house on the edge of town. It was a menagerie of stuffed animals, but one in particular had been made immortal for Hunter – a stuffed porcupine which I don't think he ever named.

During this visit we met Don Delise, an Aspen realtor who owned and rode horses in polo games. He wanted to show us his spread and watch a game or two to get the feel of the finer points. I produced many pictures and particularly a set of polo prints of the interactive nature of horses and riders. We spent a lot of time around the field and I had strong ideas for characters and horses. Less than half the pictures were used, since the raunchy story only got to Part 1. For reasons I will never entirely understand, Part 2 got tangled up with Hunter's visit to the Denver Stockyards, where horses are bought and sold. That was okay but suddenly he was raving about toxic drums of a mysterious liquid he had located in a warehouse in Denver that must be exposed at all costs as some heinous scam — 'a story of living shit', he called it. It was the last time he ever mentioned the story or horses to me.

I had also said I would help him prepare to shoot some bottles of coloured ink onto thick paper backgrounds, producing some violent art for posterity. This grabbed his attention and we set up a huge 6x4 foot backdrop of plywood and hung king-sized pieces of paper and acetate sheets on it as a target area. I hung the bottles of inks on string at different heights and let him take pot shots at each of them, or two or three at once, then watched the atomized ink drip down the sheets making its own amazing, dynamic shapes, mixing and oozing as it went. It was hypnotic and maybe a tad too hypnotic for Hunter who had by this time become involved in the sheer aesthetic of the process.

The gun he was using was a .410 double-barrel shotgun and, having now become totally involved and wanting to arrest the liberal flow of the ink after firing, he leapt forward, threw the gun to the floor and tried to hold the paper up to prevent it dribbling further. But he had forgotten that there was still a cartridge up the spout and the sudden contact with the ground triggered the firing pin and the gun went off, putting a collection of shot holes through the tank of his John Deere tractor standing nearby. As luck would have it — and luck was a constant theme throughout his life — the tractor was run on diesel fuel and was not, therefore, as combustible as gasoline, which would most certainly have exploded on impact when the red-hot pellets hit the tank. It was the one moment Hunter showed deep shame and he made profuse apologies to me for what could have been the end of me and maybe both of us.

'I always knew you were out to get me, but that was pretty crude!' I said.

'Yes, Ralph,' he said. 'I am,' he smirked, 'but not that way.

That was a sloppy mistake and I never make mistakes with guns.'

'You damn near killed me!' I barked.

'I know,' he replied laconically, 'but I failed!'

It put the mockers on the art-shoot for the afternoon, but that evening, as though to make amends and reassert his proud authority with weapons, he organized a show. Anna went inside and watched from there as Hunter set a propane bottle on a log pillar about fifty yards from the house, sat a blow-up doll in the seat of the tractor for theatrical effect, set up his Canon video camera to capture the scene and carefully took aim. He hit the bottle and, like a bomb, it burst into a huge fireball. It was a *tour de force* and gave the evening an excuse for a party.

The next day we were off to Denver and the Buddhist wedding of Juan F. Thompson to Jennifer K. Winkel up the mountain at The Gold Hill Inn outside Boulder, Colorado.

*

On the morning of the wedding, the only foreseeable problem was getting Hunter out of bed, a hungover, sick Hunter who wanted no part of it. Several old friends had gathered, including Semmis, Dibble, Gerry Goldstein and Laila, his girlfriend from the *Curse of Lono* days, and my old 'sushi partner'. She used to drive us into Aspen to frequent the Japanese restaurant when she was learning to drive.

Deborah Fuller had gone ahead the day before to arrange the finer details of the wedding ceremony and so it fell to us to get Hunter to the event on time. It was like trying to rouse a disorientated moose with its leg in a bear trap and I never ever want to try it again. Pulling the clothes off the bed, we were told to get out, and received a torrent of verbal abuse and dark curses. He needed to be sick and locked himself in the bathroom. Vitamin pills and stimulants, mystic medication were sent in, grapefruits, Bloody Marys, Dunhill's, pep pills, uppers, downers, and the Nivea Creme to lubricate his throat, from which emanated the roar of a beast in pain. We, the idiots who responded like circus dogs, had to ignore all that and persevered, doing our best to avoid his nasty teeth.

He knew he had to get up on this special day and not to make it would have meant unutterable shame for ever.

When he eventually emerged he looked like one of the living dead who has just emerged from make-up to shoot a scene from *It Came from a Swamp in Des Moins.* He was sweating profusely and guttering noises escaped from the hole on the front of his face. Then he demanded coffee, stoked his marijuana pipe, fumbled in one of his substance drawers for cocaine, got himself some ice from the fridge and poured himself a hefty slug of Chivas Regal from a quart jar, which always stood next to the kitchen sink.

We had been at Owl Farm for two hours that morning and we still had to make it to the small airfield and the waiting private jet. Wedding gear was tumbled into the cars outside and we roared off. Nine of us finally assembled beside the jet's steps. Hunter was now quiet and climbed on board with the rest of us. I sat next to the pilot and turned to look at Hunter, who was reading the day's *USA Today* newspaper and holding the ceiling above his head with his right hand. Then he fumbled with his cocaine grinder (exactly the same kind my mother used to use for her bronchial complaint. 'You didn't know your mother was a dope fiend, did you Ralph?') and snorted it snuff-style before passing it back to Anna, who passed and handed it on to the others behind her. The flight over the Rockies in a regular passenger plane of Rocky Mountain Airways was, as ever, spectacular. From the co-pilot's seat of a private jet, the mountain mass was monumental. I looked thoughtfully at the joystick between my knees and had one of those Mad Max moments, but restrained an evil impulse. I looked away out of the side window and stopped thinking.

Deborah was there to meet us in Boulder and we were ferried to Broker's Inn, where Hunter and Laila would be staying. Naturally, we made immediately for the bar — there are times when only a stiff medicinal will do. Hunter, still quiet and looking pale, had more whisky. Then, looking paler still, he was taken by Laila up to his room, where he fell into a deep torpor until the next morning.

Anna and I checked into a different hotel called the Clarion, took a swim and then went out looking for wedding gear. We did well and decided to have a meal in a Mexican restaurant. They seated us on a patio outside where, naturally, we thought we could smoke. Traffic was passing every other second and our air was a grade I example of carbon pollution. So we lit up, only to be told that even on the patio it was strictly no smoking. Americans do not understand logic and, to be fair, neither do Mexicans trying to run an eating establishment in a foreign land. The deep ingrain of codswallop infects every brain cell with the mechanical thought process of an automaton.

It was Saturday, 6 August, and Hunter's son, Juan, was marrying a vivacious girl from Boulder called Jennifer K. Winkel. Laila had ordered a white limo for eleven o'clock and was still trying to help Hunter get ready and be on time – a nervous replay of the day before. He certainly looked better and better than anything was better. She was chivvying him to get out of his dressing gown and get in the shower. He needed another drink and then he wanted room service again, mumbling and growling his familiar objections, playing for time. However, Laila, being Laila, stopped all that and finally got him into his apparel. Wearing a white tuxedo, black pants and an open-necked shirt with neck-chain and pendant droplet, he looked cool and very proud, though slightly hen-pecked. Maybe proud of being hen-pecked. Who knows what went on in that febrile brain? Slipping on his Converse Low sneakers as always and lacing them up, he was nearly ready. All he had to do was pick up his drink, take some more ice, a glug of whisky and – oh! – another extra packet of Dunhill filter tips. One more 'Move it, Hunner!' from Laila and we got out the room and into the lift. She was cajoling him again and must have said: 'Move it, Hunner!' more times than a cockatoo that has just learned a new word. She was glorious in a white satin trouser suit, resembling a very young Katharine Hepburn. Anna, too, was looking gorgeous in white, a round collar framing a gold necklace she had bought years before in Antalya, Turkey. She was also wearing the rhinestone coat she had bought in Nashville. Then we were in the limo and off up the mountain to The Gold Hill Inn. Pot, coke and whisky were consumed in rotation, and then Hunter produced the Nivea again as if from nowhere and ladled some into his mouth. Even Laila seemed surprised and said: 'Hunner! Why are you eating hand-cream?' 'Because it helps my throat,' came the reply. A bemused frown crossed her face as though it was something she hadn't witnessed before.

The limo climbed through wooded roads scattered with huge roadside boulders befitting the name of the place, past some bends and swerves, and finally arrived in the main street of Gold Hill, a dirt-track square surrounded by a straggle of wooden houses and mobile homes and the verandah-fronted inn with an open white marquee to its left. Friends and other guests were waiting and miraculously we had made it with minutes to spare. Hunter's brother, Davison, was there and Bob Braudis, the Sheriff of Aspen, just to make sure all went like clockwork.

Juan's mother, Sandy, was wearing an orange kaftan and giving kisses and hugs all round. Juan looked slim and tall, but still boyish with a white bow-tie. Jennifer was beautiful and slender in white and wearing a natural garland of wild flowers on her head. Under the marquee we took our seats and watched the bride and groom take their place either side of a flower-garlanded pillar behind which stood an Indian lady in a white sari holding a book. A bearded man in a loose, white suit sat holding a guitar, and when the ceremony began, he played a Sanskrit chant giving blessings of purity and simplicity, uniting these two happy people. Then Hunter had to rise and join his ex-wife Sandy, create a circle with the couple to pledge, in whispered tones, simple words and vows of the present and of the future and of some distant past which unites all of us and the two of them as a gift to each other. The simplicity was the very essence of a true pledge, a moment in the shade from the warm sun, a human breath of pleasure that says: 'I do.' The parents of the bride also fulfilled the act of communal promise and Juan and Jennifer kissed and they were married.

The reception was a hoot, of course. People had far too much to drink and they do so disgrace themselves if they can't hold it. Dear me! *Ad hoc* speeches were delivered *ad hoc* by wandering drunks who had already spoken when they were still in control. It was more a Buddies' event than a Buddhist one. My speech was long, but wise and focused, of course. I felt the need to impart some advice to the young couple. I am sure that Juan inherited his gentle nature from his earth-mother, Sandy. I have observed him on many occasions throughout the years looking for chips off the old block that was his father, but there is hardly a splinter to be found anywhere, except, perhaps, in the faraway look and the grumbler's mumble.

Hunter and his entourage had to fly back to Aspen. So, we all drove back down the hill for another parting drink at the Broker's Inn. Getting Hunter back in the limo to get to the airport was like a clip from the Laurel and Hardy movie *The Music Box*, where they are trying to get a piano upstairs. We waved everyone off and returned to the Clarion Hotel, had another drink in the bar and crawled up to our room. I fell asleep in front of a Kurt Russell movie and Anna wrote her diary but we climbed into bed quite early to get some sleep for our journey back home the next day.

Around midnight the phone rang and a strange voice claiming that he was the house detective said he believed we were smoking in our room. Anna took the call and protested that we were in a smoking room. She knew this because there were ashtrays on the bedside tables. He demanded to speak with Mr Steadman and I realized it was a last-ditch joke to get us down to the bar again. I was too far gone and called the operator to hold all calls.

It was the kind of joke a drunk and jovial Sheriff of Aspen would play, but it was time to go home.

WILLIAM BURROUGHS –
AN ENCOUNTER
Lawrence, Kansas, May 1995

Meeting a real gentleman . . . Shooting my art . . .
Celebrating the pain of twenty-five years . . . Giving
my all . . . Meeting the Beat Gang . . . More fireworks
and explosive art . . . Signing vintage prints . . . Giving
up on fair play

I was certain that Hunter had met William Burroughs at some time somewhere along the way, though each had denied an interest in the other and they had gone their separate ways. Hunter rarely, if ever, mentioned the Beat generation and yet, in many ways, he reflected much of their angst and roaring disdain for anything straight. He decided, ultimately, that they were different to him and he could find no real common ground.

Maybe a fascination with the works of Jack Kerouac had held him in thrall. In fact in *Volume 1* of his collected letters from 1955 to 1967, published in 1997, Hunter pays resonant lip service to Kerouac's classic, *On the Road*, by calling his book *The Proud Highway* and I find the connecting echo irresistible. Kerouac's untimely death in 1969 would have acted like a trigger, projecting Hunter into his own time-frame and any interest he had in the work of the Beat poets would have waned.

Hunter and William had, in fact, arranged to meet. Hunter was to make the journey to the little provincial town of Lawrence in Kansas but failed to turn up. Two days after the time arranged for the meeting, he had still not shown up. William had waited but, being the gentleman he was, he found such behaviour unacceptable. I do know, however, that they did exchange guns, but a chat about that with William during our visit produced only grunts and empty gazes.

We were on our way to Denver to attend my show *Making a Mark* at the Bill Havu Gallery, One on One, which was situated in the gorgeously named Wazee Street, across the road from our most favourite hotel in the world, the Oxford, downtown near Union Station. An art dealer had arranged our trip so that we could visit Burroughs on the way, to organize an 'Artshoot' together. I would do a double portrait of William, which would ultimately become a double portrait screen-print with a roundel target between the portraits. In fact, I did a 360-degree portrait using a Polaroid camera which produces manipulative prints that I call Paranoids.

I had first met William some years before at a book signing he gave at Bernard Stone's bookshop in London. In the seventies, Bernard's Turret Bookshop had became a home-from-home alternative venue for a number of Beat writers who appreciated it as a decent watering-hole not unlike their acknowledged original, City Lights Bookshop, in San Francisco. I was there most days back then and met the living embodiment of the remaining set from time to time – I considered those times both fantastic and privileged.

There was Lawrence Ferlinghetti, Allen Ginsberg in blue-denim dungarees, Gregory Corso, Terry Southern and Carolyn, the widow of Neil Cassady, who had died in 1968 of too much booze and drugs. I was so proud to see Carolyn at my book-signings in the shop. Carolyn, with whom Jack Kerouac had wanted to settle down. The shop was alive with these historic bridges to the past and I never took it for granted. Barry Miles, the underground magazine proprietor, who ran the Indica Bookshop in Duke Street, was there to record such memories and helped Brion Gysin, William's long-time friend, to arrange and catalogue a vast trunk of Burroughs's archive for possible sale to a university. It eventually went to a billionaire in Liechtenstein and money changed hands like in a drug deal.

I remember Patti Smith, the singer and poet. She came by with her friend, the photographer Joe Malanga, who took long-exposure pictures in the dark – leaving the camera shutter open for hours at a time – that produced very subtle, but, nevertheless, 'light-etched' pictures of definable shapes. I found that to be a fascinating idea on a par with the experiments in the twenties of the Dada photographer Man Ray. These luminaries were rubbing shoulders with Lawrence Durrell, Alfred Perles, Laurie Lee, Christopher Logue, Charles Causley, George Macbeth, Brian Patten, Adrian Henri, Adrian Mitchell, Henry Miller, Sylvia Plath, Ted Hughes, Stephen Spender, Seamus Heaney, Carol Ann Duffy, Harry Fainlight, Ruth Fainlight, Alan Sillitoe, Eddie Linden, Peter Porter, John Minihan (who had taken many fine photo portraits of Burroughs and Francis Bacon, who also came to the shop) and many others.

Over the forty years that Bernard Stone watched over this quintessentially supreme crossroads, everyone passed by. For some years the life-like figure of Sigmund Freud, fashioned from wax by sculptor Lyn Kramer, stood as sentinel and silent watcher of secret celebrities who may have come and gone unrecognized.

Sometime later, I met Burroughs at the October Gallery, also in London, where his gun-blasted door art had been exhibited. Frankly, I was intimidated by the power of his intellect and the frightening range of his experiences. They clung to him like burrs on a tweed suit and with those suits he wore, and that Fedora, he reminded me of a seedy small-town banker who wasn't after your money so much as your soul. His reptilian eyes and ink-line features flicked restlessly, looking for signs of life and, with unfailing courtesy, a place to enter.

I could never pretend to have become an intimate friend of Burroughs. Those claims belong to the shadows of his past that now speak to us through the words he left behind and the black and white photographs of all those buddies doing whatever they did to ease the pain. But I sensed a mutual rapport that paved the way for the eventual invitation to visit him in Lawrence.

His place was modest, a weatherboard house with a front porch in a suburban garden. The next-door house was the office of William Burroughs Communications Inc., run by what some would call his acolytes but what I would call his vital maintenance team. James Grauerholz, a musician, William's secretary and friend since 1973, had given him an ultimatum to leave New York and its ruinous influences, to stop rolling drunks for cash to feed his drug habit and to accept a controlled detox programme in the middle-of-nowhere under James's gentle protection or he would die. James's influence was that strong.

When I arrived, there were several people in attendance. The front living room was sparsely furnished on bare wooden boards. William was hunched over a walking stick – a stick with a sword in it – in a low easy chair facing the synthetic model of a Mugwump, a character from the film *Naked Lunch*, a skeletal figure strapped and chained to a director's chair. William rose graciously to view a collection of prints I had brought along for the exhibition in Denver. Brad, a stocky individual, hovered between kitchen and living room as he prepared supper for us.

William was drinking from a regularly replenished tumbler (from 3 to 6 p.m. daily) of Pepsi, which was, in fact, laced with vodka 'because I can't stand the taste of raw liquor on my tongue'.

Before the afternoon was out, we had sung daft songs to each other, discussed far more than I imagined we would and arranged a shooting session at a friend's farm the next day. He read me a poem of his called 'Pantapon Rose' and signed a copy to me. He caressed a small handgun, taking it in and out of its leather case and explaining the virtues and intricacies of what appeared to be a villainous piece of exquisite engineering. He carried it with him everywhere. He talked about cats, guns and lemurs.

We discussed the poetry sessions he had recorded for my friend the record producer Hal Willner, which had been set to music; guns in general; his fascination with guns and Hunter's similar preoccupations; the abuse of drugs and, strangely, his advocacy of respect for addictive substances. We talked about art and its relation to writing and horses, which he used to ride bareback. He believed that horses instinctively hate people and described how they can become hysterical in monumental places like the Grand Canyon, which is why mules are used there. We kept him up later than his usual eight-thirty curfew.

The 'shooting range' was a lush area of grassland, set below a steep bank beyond his friend's house and we arrived in what looked like a bank-raid getaway van, particularly when you considered the artillery we had on board. A picnic table was erected and all the guns and ammunition assembled on it. Then a silly picture was taken of William flanked on either side by me and Joe Petro, who had arrived with new prints the day before. We all held guns menacingly and posed as the Burroughs Gang. When Hunter saw it some time later, he mockingly put me on trial in the Owl Farm kitchen for bullying this 'poor old man'.

A simple wooden frame had been erected upon which our targets were nailed. The targets we had brought with us consisted of various prints of mine, including *Vintage Dr. Gonzo*, which commemorated twenty-five years of my association with Hunter, a William Shakespeare silkscreen and photo-poem prints of Burroughs photographed through a mirror by Allen Ginsberg. There was also a handsome silkscreen print of a younger Hunter, based on a photograph taken of him for his Sheriff of Aspen campaign in 1970. There were three bull's-eyes on that last print, one between his eyes, one on his mock Sheriff's badge and the last on the Rolex watch on his wrist. William was wearing a military jacket, a khaki baseball hat and blue jeans.

We blasted the hell out of most of the art and kept the Sheriff till last. When William aimed, he stood about six feet from the target. With a burst from a twelve-bore shotgun, he burnt away the best part of a Dr Gonzo image with the heat alone. Throughout the proceedings, he wandered about the range waving a selection of guns, which occasionally had us all ducking, weaving and running for cover, avoiding the direction in which the barrel was pointing. Being a gentleman, he always apologized.

During a lull in the shooting I decided to make a 360-degree 'Paranoid' portrait of William seated entirely alone at the side of the range. I moved around him, photographing his head and shoulders at specific points within the circumference. This would become the basis of my print, the 'target portrait', that would give William the opportunity to participate in his own distinctive way.

When the moment came to shoot the last print, the Sheriff, with its three bull's-eyes, I took William close to the target and explained the three areas to aim for. 'No problem,' he growled as he checked the chambers of his .44 Smith and Wesson Special (limited edition) revolver. William was enjoying himself and we felt the sense of occasion. We stood back and waited as he took aim. Then in quick succession – bang! bang! bang! bang! bang! BANG! 'Gotcha!' he said, in the silence that followed. I moved past him to inspect the result. 'You missed, William,' I said, 'they've all gone through his neck.' With a half-smile and a sidelong glance he growled: 'Well, he's dead, ain't he?'

I wasn't there for the subsequent 'Artshoot' of the silkscreen print itself. I couldn't stand the savagery or the noise and particularly the desecration of my silkscreen double portrait of William Burroughs on fine mould-made paper. But since it was my idea I had to let it happen. I had to let William blow holes in every one of the 120 prints in blocks of ten at a time with whatever weapons came to hand. That was his contribution to our artistic collaboration and I wanted him to have fun. I asked him if he would like to give the print a title, lending his contribution a literary dimension. He thought about it for some time, some days, in fact. I received a phone call from him when I had returned home. His title was: *And Something New Has Been Added.* Irony!

Then we were off to Denver and Aspen where we had arranged to have a twenty-fifth anniversary celebration of the birth of Gonzo. Hunter would sign the *Vintage Dr. Gonzo* print and the *Shot Sheriff* that Joe was driving all the way from Lexington with his girlfriend Polky – 'for safety'. Prints need to be kept flat and don't take kindly to rolling.

Anna and I had, by this time, gone to the home of Gerry Goldstein, where we would be staying for the visit. Hunter arrived and made himself at home, raiding the fridge, getting himself some drinks and demonstrating his prowess as a chef, which he had never done before, for me, at any rate, in his own place. He made us a stack of mustard and ketchup sandwiches which served to scour our throats and alleviate the hunger that would be completely assuaged when the girls got back from the supermarket. That was a start but Hunter wanted to get back home to watch a basketball game and he wasn't going to sign just then, anyway.

We received several calls from Joe and Polky and it was obvious that they were lost but could find the Woody Creek Tavern. So, we agreed to meet there. It had been a long drive for them. The snow was still on the ground and they arrived looking like Eskimos lost on the ice.

When we arrived at the Tavern, at some ungodly hour, the arse-end of a party was just running down. Even the staff were in party mood and someone asked Anna, in a phrase that became a catchphrase for evermore: 'Annie, wanna party?' It has been used ever since, whenever someone gets out the drinks. We were going to drive back to Goldstein's house but were stopped by a patrol car en route. I was asked to explain why I was driving without lights. The traffic cop would not believe that I had been drinking. I had borrowed the red shark Chevy and was obviously in control. It was just that I couldn't find out how to turn on the lights. Anna's cut-glass English accent sorted them out and they even obliged by showing me that the lights were somewhere near the pedals at my feet. I laughed an uneasy laugh and found them. From there into town we had a police escort all the way to Gerry's house, turning right at the familiar Jerome Hotel. It is amazing how sober one can be when one has to be.

The next day, 6 May, was the twenty-fifth anniversary of the birth of Gonzo. It was exactly twenty-five years since Hunter and I had met. It had to be special and, to celebrate, we planned to watch the Kentucky Derby on TV and everyone had to bet.

We were late getting there and I failed to back my choice, Thunder Gulch, which, of course, romped home at 20–1. I always win when there's nothing at stake. Hunter was actually up and dressed, which was odd, and Joe had the *Vintage Dr. Gonzo* and the *Sheriff* prints in the back of his Trouper truck, hoping to catch him during the day and get the signing done. There was the bonus of Hunter getting out some guns and blasting away at some of the out-of-series prints, to make it more interesting.

A big outside table was used to lay out the guns, cameras, ammunition and skin cream and to hold the drinks. The sun was out and the snow had more or less melted away. Across the valley, on the ridge, the aspen trees were still without leaves and displayed a moody colour of burnt umber in the shadows. A tracery of trees, nearer to us, drew lines of ochre through the background like crochet.

It was turning into a good day. Joe gave Hunter one of his metal sculptures and that became the first target. 'Oh, thanks!' he said. 'You're a kind man, Joe.' He walked about twenty-five yards away, set the sculpture down and walked back, turned, took aim and hit it first time. It bent into a fascinating twisted shadow of its former self. The peacocks on the verandah were undeterred by the bangs, as were Anna and Polky, who had decided to become spectators, too, and talked furtively together. We had the requisite mint juleps and an anniversary iced cake with striped black and white icing, and strawberries. The words 'RALPH + HUNTER. DERBY DAY. HAPPY 25TH' were written on the icing in chocolate. Joe produced the prints, determined not to let such a relaxed occasion pass without getting the signatures. The flat box was laid out on the trunk of Hunter's gold Cadillac and the signing began in earnest. Hunter got to work with a will. I noticed that once he had begun, nothing stopped the flow and all was signed with a furious determination. The trick was to get him to that point. A motor seemed to start up inside him and off he went. You had to read the signs. He didn't want to do the 500 *Vintage Dr. Gonzo* prints then and said gruffly: 'Later!' He would do them that night at Gerry Goldstein's house.

Deborah had prepared a fantastic meal of salads and prawns, rice and fruit. And Maggie, who typed for Hunter, lit a fire in the open grate in the living room. The afternoon wore on and we moved inside for a while to eat at the low round table and, as often happens, Hunter produced some of his poems and got me to read them, incoherently, but with no little panache, of course, to the assembly. Hunter listened carefully and moved his hands like an orchestra conductor. However, the Philistines started talking amongst themselves and so I delivered some of them in a W. C. Fields voice. Hunter hated that. I shouldn't mock his work.

After that it was everybody outside for his famous 'bomb' trick — the ritual shooting of a propane gas cylinder with a high-velocity rifle. The John Deere tractor was brought around with lights blazing, the blow-up doll with the voluminous tits was sat in the driving seat and the engine was left running. Her boobs bobbed up and down and Hunter filmed this one again, but this time in the dark. The fireball erupted, sending metal flying around like shrapnel. The bulk of the cylinder was found the next morning on the other side of the field. As luck would have it, nobody got hurt but all were now fired up to cut the cake.

It was one in the morning when we left. Joe drove me in Deborah's daughter's clapped-out truck and Polky drove Anna in Joe's car. She kept flashing her lights at us as Joe was driving far too fast in a 25mph zone and, sure enough, the Sheriff's car was on us but stopped Polky first. She simply charmed the officer with her strong Southern drawl and we were on our way, once again, with a caution.

Hunter didn't show up that night but turned up around ten the following night with Maggie. We had eaten and were considering an early night, turning out the bank of lights which Americans love to shine into the tiniest corners. Both ate some *coq au vin* which Anna had made earlier and then the table was cleared to make way for the huge edition of *Vintage Dr. Gonzo*. Hunter didn't flinch, immediately starting to sign, surrounded by Joe, Polky, Maggie and Anna. I watched him do the first twenty-five prints and then lay down on the

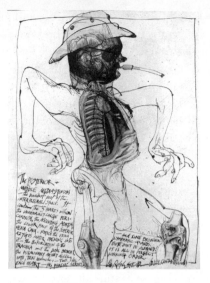

sofa. I was awoken by clapping and a cheer. Hunter had signed them all with the help of Polky's easy Southern cajoling. She had eased him along and even got a smile from time to time, as she helped to stack the prints neatly as each was signed. Someone had thrown a blanket over me and I watched from the sofa. I couldn't believe it. It had been done and I hadn't had to lift a finger.

Joe made off the very next morning through the snow that had fallen in the night. I was trying to get Hunter to give me some words for *Polo is My Life*, but he wanted to watch another ball-game he had put some money on. Houston versus Phoenix. I don't know who won. Why should I? Hunter knew nothing of English croquet, 'the sport of decent folk', I reminded him. He claimed that he betted to fund his drug habit. It was as good an excuse as any, I guess.

On this trip, Hunter was more insistent than on most for me to read some poetry by the fifteen-year-old daughter of a mother involved in a weird supernatural magical cult. He read some of it himself, but the timbre of my Welsh lilt touched the marrow of this girl's beetle juice. Sometimes, he would crusade on behalf of someone he had just met at the Woody Creek Tavern. There would be no sane reason why he should but it always sounded vital and of critical importance. This was another and I went along with it, for in my way I considered that what was good enough for Hunter was good enough for me. On this occasion, however, he started to write something about me. I was a necrophiliac, I was a stalker. I 'stole' photo pictures of him and drained his blood like a vampire. All the friends who arrived with me were feeding off him, paranoia was his friend and nobody was beyond suspicion. He worried about people taking pictures of him as he reached inside his drawers and cabinets, grinding coke and snorting liberally. He savaged my attempts to be discreet and he berated me, denouncing me as a spy who had long since exhausted my role as his partner and who could now do wrong.

Everybody put up with this, some for a quiet life and some because they loved being beaten on by the holistic love/hate relationship Hunter had with life and with those who peopled his life. Some were more philosophic than others and, if they didn't turn a blind eye to his raving and his ranting, took note of it as one would observe a caged lion, quietly making mental notes and dining out on it as something of interest later. In some ways he *was* in a cage. He had put himself there quite wilfully, declaring Owl Farm to be his 'fortified compound' and living off it for damn near forty years. In that time he

made friends, acquired them, seduced them and made them all part of a coterie of other animals and they each had a role. Some already had plumage, feathers or fur, teeth or claws or both. Some were famous and some were unwitting neighbours who became his friends. But no one lost out unless they came wanting nothing but a story. They did not come in friendship, but came to flatter and take.

After all this we had a show in Denver and took the Rocky Mountain airline down to five thousand feet to the brand new airport, a Barnum's Big Top type of a place, but fifteen times as huge. The gallery on Wazee Street is in the old downtown area, a one-time cattle centre, which is now a complex of renovated warehouses, providing restaurants, galleries, boutiques, men's hairdressers and bookstores. It was just getting into its stride in the mid-nineties and wondering what it was going to be. It is now known as LoDo, which is short for lower downtown and has a free, on-off, hybrid bus service shuttling back and forth up the centre of town on 15th Street.

The *Making a Mark* exhibition was crowded on the opening night. All the locals who knew Bill Havu and their Mayor, John Hickenlooper, were there.

We had hunkered down in the Oxford Hotel again, mercilessly using McCormick's Bar next door to do interviews. It was then I wondered how I would remember all these folks in later life or were they just ships passing in the night. Hunter didn't show yet again but, by this time, I knew that if he wasn't going to be the centre of attention, he never showed up – not even for me. He didn't show on two other auspicious occasions which I never forgot, British Design Week in Aspen, in 1986, when such luminaries as Norman Parkinson, Zandra Rhodes, David Hockney, Saul Bass, Milton Glazer, James Stirling the architect, and a host of others were present. All were easy to talk with but they weren't there to see Hunter – necessarily. However, he would have been mobbed, if he had only realized. The other notable occasion was my show in Aspen, at Barney Wyckoff's gallery. Many came along then but Hunter was nowhere to be seen. Maybe I should have inveigled him into the place with works of his own and made it a group show, because he would have been there with an excuse. I didn't realize it then but that was the underlying reason. Basically, right up to the end, in spite of all the big stars and high rollers, he was the only real show in town. He did art, tried it and made it work his way, which was subconsciously to contrive it from those people he had a pathological desire to blast with guns, eliminate, and, in the same way, display his contempt for them. I never realized until now. The last thing on my mind had been to steal any of his thunder whatsoever, but, on the contrary, welcome him into all of it and be proud of the fact that he was my friend at all.

RIFLE
March 1996

Touchdown in a snowstorm . . . Too many people . . .
Hunter on trial . . . Making an exhibition of myself
. . . High jinks at Owl Farm

Rifle. Now there's a name to conjure with. It sounded like something a wagonload of pioneers would dream up. It was 5 March 1996, but it was still winter in the Rockies.

Our plane had to land in Rifle in a snowstorm. I was on my way, with Anna, to Aspen for a one-man show of my work at the Barney Wyckoff gallery. To some people snow is fun but to me it gets in the way of walking. You are compelled to wear big boots and heavy waterproof clothing and walk like Boris Karloff in *Frankenstein*. People pay small fortunes for the dubious opportunity to kit out entire families in primary-coloured one-piece zip suits made from slimy nylon. *En masse*, they remind me of fashion-conscious penguins in an ice cream parlour. There are massive real estate plans for the area and an enlarged airport development, waste-water projects and God knows how many overpriced new homes on the outskirts for those who just can't make it in Aspen. Oil shale leases are being proffered and reports of new departments within the city's ambitious development plans will provide employment for those who will file reports about environmental impact on the region, thus enabling new hordes to live there and perpetu-ate the cycle which ploughs on relentlessly and effectively fucks it up like everything else. I guess we all have to shit somewhere but human beings seem to need to shit more than most. Rifle, once small, quiet and remote, is flexing its muscles and doing its bit to transform the Rockies into a heaving mass of scrambling humanity.

We were put on a bus to Aspen. Signs of this new development were beginning to blister and erupt erroneously in brand-new ice-cream-coloured estates, with malls, banks, garages, shops, town halls and of course Wal-Marts at every bend. Holy God! We were strangers in a familiar global village and there was nowhere to hide. The Rockies were fast becoming everyone's place to reproduce and move on. One day those coming from the east will come face to face with those coming from the west. Then one of Hunter's worst nightmares will be a reality. There will be nowhere else to go. There will be one hell of a fight and survivors in the not-too-distant future will start over and make exactly the same mistakes again on a pile of polluted rubble.

Anyway, we got to Aspen and nothing seemed different except that a burgeoning sense of opulence prevailed. Aspen was now too smart for its own good and polo and skiing were the local sports and spotting celebrities who had bought into this white Eldorado was a non-smoking coffee break pastime. We were staying at a hotel called the Prospector Condominium, as though we were here to mine for gold or even silver. The condo was directly opposite Barney's gallery. It's all here, after all.

'How d'ya like it so far?' said Barney, who picked us up from the drop-off point at the airport.

'Rifle's fucked,' I said drowsily. 'When are you going to let things be?'

Barney drove us to the condo and saw us into our room, which sported an outside hot tub. 'That's for me!' I thought, 'but first let's sample this very fine, duty-free, single malt, The Balvenie.' Then I got into the hot tub and contemplated. I would phone Hunter after that and invite him over. A condo, I had just found out, is a place where you bring in your own food and cater for yourself. So we were not in a sweat. No rushing around trying to find an eating rendezvous.

We waited for Hunter and he arrived around midnight, preceded by hoots and shouts and a bullhorn commanding us to come out peacefully with our hands in the air. He was carrying a bag of 'essential equipment', torches, retractable knives, a pipe, dope, his own unusually small bottle of Chivas

Regal as well as a plastic hammer. He had brought along a tiny girl called Susan who lived in Snowmass, a ski place back towards Rifle.

'Have you brought my drawings, Ralph? *Polo!* We have to finish the story. I need your drawings!' I had brought a few horse-related drawings on the off-chance and we looked at them like a customs officer looks at passports. It was all he would need, I reckoned, to complete the second half of *Polo is My Life*. 'We have a problem, though,' warned Hunter. 'I have discovered this warehouse chemical dump and it's not a pretty sight.'

'What's that got to do with *Polo*?' I asked.

'Fuck all!' he replied, 'but I can't ignore it. It's a horrible scar on the landscape and it's in Denver!'

'Well, maybe we should make that an entirely different story,' I reasoned.

'There is no reason in this scum, Ralph! These bastards need showing up right away.'

'Tomorrow, Hunter, tomorrow.'

I never heard any more about it. Neither did I hear any more about *Polo is My Life, Part 2*. Notes must exist but the finished story never surfaced. One day his trusty archivists will unearth the bones of the story's climax. The pictures already exist.

After looking in on the Show at Barney Wyckoff's gallery, we were due to go up to Owl Farm that evening for a party and to unwind. It was the night Hunter accused me of gross behaviour towards an old man — William Burroughs — who had posed for a photograph with me and Joe Petro as the Burroughs Gang in 1995.

It was decided to put me on trial. I filmed the 'court-room scene' in Hunter's kitchen during which I was sentenced to live out my worst nightmares. I was denounced as a bully, a mendacious publicity-seeker and a fraud. Instead of 'doing my sentence' I agreed to play the part of a woman and sing a song.

Anna and Barney went back to Aspen to the back of Barney's gallery where clothes were kept for cabaret in the bar next door. They chose well and helped me to dress up, lathering lipstick over my lips and practically over half my face. Wearing a black, velvet, sequinned top and spangly, silk, patterned dress, a blond wig and a cut-brimmed straw hat with veil, I entered the kitchen with a camp swagger, using my fully formed 'Northern Queer' voice, one that had never before been heard of in Colorado.

Gerry and Deborah manned the camera and I performed. I explained how I had first met Hunter, thought he was

beautiful and fell for him immediately. Then I went towards him in a provocative manner and slobbered lipstick all over his face. He was embarrassed but amused. Though I have often wondered if he might be a bit homophobic, he didn't show it that night and entered into the spirit of it all like a pro.

It was a bawdy performance and full of fun. We got the laughs we were looking for, anyway, and only a slight reference to the conservative politician Pat Buchanan gave Hunter pause to caution me. He didn't want to mention Pat Buchanan, for whom he seemed to have a soft spot from his years in Washington. Whether I referred to the religious right I cannot remember but somewhere inside him a raw nerve winced. Perversely, I guess he got on with him as well as he did with Richard Nixon once they had found that they had something vital in common — football; and Nixon knew his football.

There was a long alcoholic night ahead, plenty of chicken, pasta and salads, prepared by Deborah, who never scrimped on food for guests, even though Hunter as usual treated his food like a messy child. Then we got to talking about Hunter's upcoming trial for drunk-driving in which Gerry Goldstein would be acting for the defence. At the trial that summer I

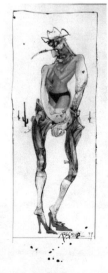

was privileged to see Gerry in action and sent him this photograph after the verdict of Not Guilty:

Though Gerry believed Hunter to be innocent, even though nobody is actually innocent, let's face it, the nature of the entrapment put the police in a very bad light. When Hunter took the stand he described how the police officers lay in wait for him on the road to Woody Creek and said that they were hiding under the bridge 'like trolls'. There was laughter in court.

It caught the essence of those who would try to mess with him. Woody Creek and its environs were Hunter's personal stamping ground and his death has left a deathly hole in the heart of the community. The trial was a formality and a foregone conclusion, although I am sure it was preying heavily on his mind that night we had the party. Even 'trying' me was a way of lightening his load.

VISIT TO VIRGINIA,
HUNTER'S MOTHER
Louisville, 1997

*A rest home in Middletown . . . The family bartenders . . .
Memories of Hunter . . . A present for Virginia*

The Kentucky Derby in 1970 was the first and last time that
Hunter introduced me to his hometown haunts and he never
introduced me to his mother, Virginia. Back then he had trouble
telling me that part of the reason he had come back to Louisville
at that time was to help his mother get institutional care for
what sounded like a drink problem. It was years later, when
Virginia was in her nineties, that I was determined to meet her.
Anna, Joe and I went to visit her at a rest home outside Louisville.

'Don't visit my mother, Ralph! I will never forgive you!'
Hunter harboured a deep fear of his friends meeting those
from his past who knew his worst faults.

The rest home was in a historic region called Middletown,
outside Louisville — Tom Sawyer State Park. It was bleak in
the rain. We passed endless rows of fenced-in weatherboard
houses, white churches with black crosses, deserted tennis
courts with puddles and wire enclosures, weather-bent black
figures leaning against the rain, and us splashing along towards
a tarmac entrance beyond the concrete blocks of new estates
and into the Episcopal Church Home, odd since Hunter had
no time for religion.

The home masqueraded as a hotel. Glass entrance, electronic doors, expansive foyer with louche, soft sofas for visitors.
We announced ourselves and were greeted warmly by a staff
member schooled in the art of making people feel welcome,
whatever the appearance of the visitors.

'Have you come far?'

'Further than usual. Twenty-four hours.'

'That's a long way. Are you family?'

'Hell, no! Mrs Thompson needs new insurance to beat off the opposition.'

'Opposition? What opposition?'

'We were told there was undesirable competition in town in need of her business.'

'Her business?'

'Of course. You can't be too careful. She is ninety-two years old, for god's sake. She can't go on for ever.'

'In our care she could live, maybe to . . . er . . . one hundred and two and maybe more. We have one inmate who is one hundred and seventeen and she is getting married next Thursday.'

'Have you asked her about insurance?'

'All our clients are insured, otherwise we couldn't take them on.'

'So you are in line like everyone else to catch the windfall,' I said.

'Mrs Thompson will see you now.'

A nurse in a gown of pale-blue frou-frou stood in the hall and beckoned us along the corridor behind her. We waited and wondered. She might be horrible, a chronic shadow of malevolence, railing against those who would disturb her savagely earned peace.

Virginia Thompson's door was one of many along a wide corridor on the ground floor. She had told me on the phone that she had arranged for a bartender to be there in the room to fix the drinks.

I knocked gently and a gravelly voice growled: 'Come in!' I pushed open the door and diametrically opposite from me, across the room, sat a tall, handsome, big-boned lady in a Lazy Boy chair with a king-sized filter tip in her hand and a freshly poured bourbon in front of her on what looked like a drinks trolley zimmerframe. There were two other women there.

'Welcome!' she said. 'Meet my cousin Virginia and her daughter . . . ah . . . Virginia. They are my bartenders.'

Finding it difficult to move I leaned forward and kissed

345

her on both cheeks, which she remarked upon. 'Huh! Two for the price of one,' she smiled, not looking at me, in exactly the same way as Hunter would remark about such a greeting, looking into the middle distance as he waited for a response.

I would usually keep absolutely silent, absolutely quiet and look in the opposite direction, walk through into Hunter's sitting room and make a re-entrance before flamboyantly hugging him like a long-lost homosexual friend, using my effeminate Northern queer voice.

Many pictures, mainly family photos, were hung around the walls behind her. Deborah Fuller had been to visit a couple of weeks earlier and had put them up for her. There was one of Virginia in her wedding dress with her husband, Jack R. Thompson, an insurance agent. Hunter, Davison and Jim, the youngest brother, who died of Aids, were in another photo as boys, in short-sleeved pullovers like the ones my own mother used to knit for me and the regulation grey flannel short pants.

Virginia spoke warmly of the time I first met Hunter back in 1970. He had talked to her about our first assignment. He had been concerned about her drinking at the time, but, she said, 'Hell! I was just as concerned about his. He was the one who needed the straitjacket, not me! He finally got me into one. I don't like it but I guess it's better than the gutter. I had to accept it.' As we talked, I scanned the room and my eyes fell upon an early painting of Hunter and Davison, looking gentle and cherubic, probably done by one of those professional family portraitists whose policy was always to flatter. Davison was holding a model aeroplane and Hunter was clasping a book in both hands.

I liked Virginia a lot and remember thinking that I wished I had known her earlier.

The drinks were liberally administered by the other two Virginias as though they had been instructed not to let the side down. I did a drawing of Virginia and we stayed another half-hour after the 'bartenders' had left, as though their task was complete and a suitable impression had been created. Thankfully, Joe had been drinking cola as he was the driver.

Virginia liked weird art and so we decided to make a second

visit and present her with a huge colour silkscreen print of the *Lizard Lounge* from *Fear and Loathing in Las Vegas* that Joe and I had just produced back in Lexington. Picking it up the following day from the framer's, we could see how magnificently rich and vibrant it looked. It had been framed in Perspex and was therefore quite light to carry to and from the car in spite of its size.

We found Virginia sitting in her wheelchair in an alcove of the lobby with a shrunken and smiling old lady, her buddy from across the hallway. She had been waiting for us with some anticipation, she told us, and wheeled herself back into her room in our wake.

Once again Virginia insisted on extending the hospitality of the previous visit and cajoled me into having a glass of wine, while Anna was dispatched to fetch her a Coca-Cola from her little fridge and a pack of cigarettes from her bathroom cupboard.

Virginia's response to the print was joyous.

'I love the colours,' she said, 'though I don't understand what it's about. Maybe you could explain it to me. It's going to cause a lot of trouble around here. So that's good!' she added gleefully. The print fitted perfectly into a niche between her bedroom and the door.

Our conversation centred on the visits that Hunter had made to Louisville for various lectures and appreciation ceremonies at local institutions such as the arts centre and university. Virginia described how there would always be drinks provided before these events and how on one occasion she was disgusted that wine was provided rather than the expected bourbon, since it was, after all, a party in Hunter's honour. She also described how Hunter had a girlfriend at the time whom no one liked since she was extremely possessive about Hunter, though Virginia conceded that he didn't 'treat her right'.

'Hunter used to let her out of the car and drive off, saying that he'd left her somewhere on 77th Street.'

Then Virginia's reminiscing turned to a probation officer who worked with Hunter as a teenager when he was in trouble with the police. He was supposed to be home by ten at night but he rarely was. The probation officer would sit on the porch with Virginia waiting for him to return. She still kept in touch with him, though Hunter never acknowledged his help. It seemed that he did a lot for Hunter.

We said our goodbyes to Virginia at her door but she insisted on wheeling herself to the front entrance. As she passed the enquiry desk she invited the receptionist to view her art and very soon, I learned later, all the other folks in the home had seen it too, hanging in pride of place at the entrance to her room. It was there until the day she died. It was the best thing I could have given her.

THE END

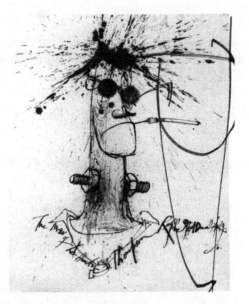

'Ah Ralph, you filthy little animal, filthy little beast, I have
a job for you, a proud and noble job.'

THE RED SHARK

2000

Breathless at the Delaware Hotel . . . Old bikers
never die . . . Fishing like pros . . . Near death on
the Basalt road . . .

The radio gurgled in room 520 – through my headphones –
into my ears, my big ears, laying spread out either side of my
head, on the pillow I called my own. Three a.m. There was a
radio phone-in on cloning. Everyone who can speak has a
crystal-clear opinion on the subject – and they were wide awake.

'Imagine we are entering Huxley's Brave New World,' said
one and he began to explain his idea. The thought grabbed
the imagination of the presenter, who grasped the nettle and
ran with it before the caller had a chance to say: 'Hey! Wait
a minute! That's my idea.'

The idea that we can, at last, have neuter beings, the workers, neither male nor female, nor thinking creatures, but obedient slaves to save the natural-born-us, to do our bidding. This is the utopia dreamed of in us all. The radio presenter droned on, grafting someone else's idea, seamlessly, onto his own. The phone-in Aristotle was lost on the night-time airwaves of non-stop chatter. The presenter had learned to fast-talk his way into the ideas of others and leach them to become his own.

Leadville's dawn light crept up a deserted main street. Not even a stray dog looking for a place to sleep. I had crossed the room of course, maybe the same room that Butch Cassidy slept in or Doc Holliday coughed in, to look out the window onto the desolate scene outside.

I could say that a mountain cat was prowling beneath the shadow of the bizarrely named Manhattan Bar, right on the next block on our side of the street that only hours before had been a heaving, sweaty waiting room of habitual drunks and heavy smokers, but it wasn't. The silent, deserted atmosphere willed the imagination to fill it. No luck. The silence was enough. Our view from the hotel room – a two-room suite with diagonal corner windows, looked straight down a sweep of road through the heart of 'Oro City', Magic City, Cloud City or just plain Leadville. It has had many names in its history, depending, I guess, on the luck of the dreamer who came looking for gold and found only a cloud of dust. Nestled in the heart of the Colorado Rockies' tallest peaks at over 10,000 feet, where the air was clear and cold, the town had seen better days. The hotel was the Delaware and it had seen better days too, but was none the worse for it. The distressed doors, décor and fittings conjured up distressed sepia images of Wild West carpet-baggers in plaid suits and toppers, passing through and looking for mugs to buy eternal youth, miracle potions and acne healers. Old photographs of prominent town dignitaries and silver barons drooped along the gloomy hallways. Groups of long-dead citizens, miners, most of them, stared out from photos taken on old wooden cameras on solid wood tripods by men in tight suits and bowler hats. They were smart enough to realize that they could make an honest living travelling across America with mobile darkrooms in which they could also sleep, capturing everything for posterity like Matthew B. Brady covering the American Civil War.

Oscar Wilde must have stayed here on his way around America delivering his lectures on the Aesthetic Movement. Although aesthetics wouldn't have meant a butt dick to the average miner, or dignitary for that matter, Oscar held his audiences at Horace Tabor's newly built Opera House in thrall with his risqué raconteur's banter and dandified appearance. Oscar came to town in the 1880s attracted by a frontier romance book tour and the new wealth society, drinking and carousing with the mining fraternity, who, by all accounts, loved his exotic behaviour and use of English. Leadville must have been a practically unreachable mining outpost in those days. Roads would have been mud paths and any physical comforts were rashly forsaken in favour of artistic pursuits and lawless abandon. The miners knew a class act when they saw one and invited Oscar to go down the mines with them where much whiskey-drinking ensued. He never stopped talking and could hold his own with the best of them. They named one of the shafts after him. Oscar's Mine.

The lobby of the hotel was a vast floor with supporting square pillars, largely unchanged, where the weekly auctions of copper, silver and lead ore were held. Every miner, whore, huckster and pickpocket would have made those days human goldmines and a preacher's paradise of lost souls and patent medicines.

It was easy to get carried away listening for the echoes of long-gone scenes. In the morning light a group of ageing Hell's Angels jerked me back to meet the new day. Suddenly it was the twentieth century again, almost the 1960s, but this wasn't Altamont and the Rolling Stones were not playing. This was a bunch of guys who loved to ride, still longing for the freedom of those times and still driving the highway looking for something that passed them going in the other direction. Nostalgia is a powerful drug and their chrome-plated chariots stood in front of the hotel in rows, reflecting their owners' snap-black gear as they emerged into the morning sunlight.

'I'm just off to visit Hunter Thompson,' I said, taking a quick group shot of them slipping on their chrome shades as they mounted to leave. 'Up the hill in Aspen. Do you know him?'

Sure they knew him. One called Greg went to school with him in Louisville. 'Tell him Hi!' They took an inordinately long time preparing to take off, revving up, adjusting gear and kicking the peg jams into horizontal. When they finally did leave, they peeled away like a squadron of fighter planes, in one long, drawn-out, majestic roar. It could have been a carefully orchestrated ballet on wheels and probably was. They were proud of their art form.

We, that is Joe, myself and Anna, took off with slightly less panache, stopping first to buy fishing lures, spinners and spiffy new collapsible rods and spinner reels. The fishing in Colorado rivers, so we were told, was second to none.

Well, it was for us. Frying Pan River must live in infamy for the serious fishermen who were there that day when we amateurs arrived. They sneered like professionals at our cheap tackle. We were not, like them, togged out in the finest gear you could buy from Orvis Outdoor Outfitters, especially the waist-high waterproof boots and weather-shaped hats that

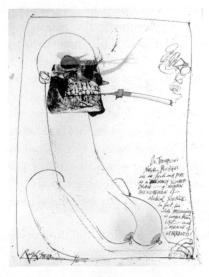

separate the committed fisherman from the stumbler. Everything matched, each with the other like a Masonic lodge of anglers. They were clones to perfection. These were men who needed everything for the titanic struggle when the big ones are on the hook. Their whiplash rods glistened in the sunlight as they cast their lines half a mile into the dark, moody waters beyond, from wherever they stood. Silent concentration and an air of absolute authority were their weapons. It was a humbling sight and we wondered whether our fishing licences had, after all, been a shameful waste of money.

Which was the real reason we had to try. We cast as though we knew a thing or two. In unison, like Butch Cassidy and the Sundance Kid, side by side, we Proud Sons of the Kentucky Pioneers, hurled our gaudy lures in every direction as though we were going to shoot the fish out of the water, which of course we did.

One after another we dragged them out. Two- and three-pounders were flapping on the grassy bank behind us. Anna opened the trunk on our Trouper and began to load them on board like a fishwife in a frenzy when the fleet comes home. We tried not to catch the glassy stares of the professionals, who by this time had stopped casting and were standing with their rods drooping in the water, forgetting to reel in and letting their lines sag and drift downstream as they watched this flagrant display of unethical entrapment and bad sportsmanship.

'Time to go!' said Anna. 'You mustn't be too greedy. There's always another season!'

We tossed our hastily wound rods into the truck and backed out to cross the bridge and head on down the towpath along the river towards Aspen. We looked back and watched the professionals. One by one, trying to look casual, they sidled towards where we had just been. They peered into the swirling waters, pretending to move on, looking at their sportsmen's waterproof watches, but one by one attempting to find those magic spots that only moments before had been a thrashing cauldron of mortal activity.

A short, tortuous climb through Independence Pass, the ball-breaker assault course for those pioneers who got so far, and we wound our way down into Aspen, home of refugee hippies and puff-nylon ski chic, where all was arranged for us to stay at a log cabin at the kind invitation of George Stranahan and his wife Patti.

And then the moment could no longer be put off – to drive up to Owl Farm, up the mountain, past the little town of Woody Creek with its trailer park, store and tavern and the steep bend to the right, past pasture and horse farms to the driveway at Owl Farm with its rusty metal vultures on guard on the tall gateposts.

There was no prospect of seeing Hunter. He was asleep and had been all day. George popped in and regaled us with local tales, including the fact that he and Hunter shared an obsession with dynamite. In the middle of the night we heard raucous animal cries . . .

Owl Farm hadn't changed one bit. Rusty vultures, open pinewood gate, peacocks strutting, constantly weathering.

We were surprised to get a phone call from Hunter at about nine o'clock, just as we were thinking of getting some break-fast together. He had been asleep for a day and a half and suggested breakfast at Basalt (a small ex-mining town that I have always liked). So we drove up to Owl Farm and into the kitchen to find Deborah and Anita, Hunter's new assistant. A nice girl and pretty. Hunter emerged from his shower, soaking wet in his bathrobe, and the usual searching in his pill drawer and constant replenishing of his drink began amidst the general conversation.

Deborah, Hunter's do-all, fix-it and faithful protector, had backed the Red Shark, a restored seventies Chevy, out of its garage, its first trip for a while. Maybe it was the two-door coupé Camaro Z/28 front engine, rear drive, 350 cubic inch V8, delivering 330bhp and a tire-vaporizing acceleration rate of 0–65mph in 6 seconds and weighing in at 3218lb . . . though I can't be sure . . . It dazzled glamorously in the high altitude light; its rear, svelte buttocks thrusting back contemptuously towards the vehicle it had just passed on the inside hard shoulder. The expansive beige/crème sofa seats warmed luxury in the sun. Set up beside the four-speed manual gear-shift was Hunter's *pièce de résistance* – a polished walnut drinks tray taken from some elegant, long-gone Rolls Royce. It was large enough to hold a wide glass of scotch on the rocks, another glass of spare ice and a bottle of Heineken.

It had been arranged that Hunter would drive Anna, Joe and me into Basalt, another old mining town, for breakfast. Hunter's treat. We were in a good mood. 'Anna. You can sit in the front next to Hunter,' I said. 'Joe and I will ride in the back.' Anna resolutely insisted on riding in the back with Joe.

As Hunter slipped the clutch and we coasted down the driveway, Joe lifted up crossed fingers. Anna grinned ruefully.

I have had one other near-death experience in similar circumstances; so I recognized the symptoms. An atmosphere of eerie calm, mingled with high spirits and the child-like anticipation and excitement of a school outing.

We began to pick up a hell of a pace, the wind whipping past us, taking the bends at a tremendous rate. Luckily, our destination was only a few miles down the road.

Basalt is now a thriving wealthy town of picturesque weather-board log cabins and tourist shopping arcades. Leafy aspens tickled the atmosphere with theatrical fingers. There is a river hooking around the town and steep-rising basalt rocks wherever you look. A table outside on the balcony of the restaurant, overlooking the river, was a perfect setting for a relaxed brunch with drinks and easy chatter. A couple of other friends turned up and we set to work demolishing four hours in what seemed like five minutes.

We were joined by Wayne Ewing, the film-maker, who had been filming Hunter for years as a progressive archive and footage for his film known as *Breakfast With Hunter* – and a slight, doll-like woman called Kathleen. Unfortunately there was no breakfast menu but Hunter kept ordering different dishes – grilled shrimps, chicken, guacamole and salsa, etc. He was drinking large tumblers of cranberry juice and gin.

Going back was when I felt the fear. There was no panic, because you know what is happening is happening precisely to YOU and you sit back to accept the events. I had a deep trust of Hunter's ability to handle a car, as you know. I don't know why I knew this was going to happen but I just did. So I didn't fix my seat belt and neither did the rest of the party. Come to think of it, I don't think there were any to fix. In fact, I don't think seat belts were standard in any Chevy, even the Corvette Stingray, until the late seventies.

Anna and Joe were sitting in the back and Anna had her eyes covered with the palms of her hands, as though she didn't want to know what was about to unfold like a clip from a film.

Hunter pulled out across the two-lane highway and surged back towards Aspen. I looked straight ahead across the landscape of the huge rolling red hood. Hunter, characteristically, had his left hand on the wheel while the other manipulated ice from one glass into another glass of whisky. It was a difficult manoeuvre at the best of times but, like me, he suffered from arthritis and arthritic joints play havoc with the motor reflexes. Neither of us could any longer drive in the fast lane the wrong way with the headlights out.

It was an exhilarating moment and we were enjoying the ride. Hunter was showing us the sheer elegant power of gliding on a suspension that disguised even the slightest bump in the road.

The sky covered with clouds and spots of rain began to fall. 'If we drive fast, we won't feel the rain,' said Hunter as we careened down the main Basalt–Aspen road.

I remember the moment. There was a black truck, a truck with a dark history of toil and dirty jobs, full of sticks. Oily sticks. It was too dirty to have carried anything but dirt. It was born to carry garbage. That was what was so weird. It was a huge block of trundling garbage in front of us, a mobile obscenity, and a crystal-clear light beyond, beckoning. Hunter instinctively put his foot down on the throttle to zoom past this monstrous obstruction. Hunter could not live in shadow.

We forged ahead, as did the truck, suddenly, and we drove side by side, swaying like wild drunks on skateboards, laughing at some distant shore.

The highway became a two-way road quite suddenly and a squadron of about six cars were coming in the opposite direction, as cars do when they are on the same road as you. They were swerving away towards the steep bank of the Roaring Fork River. The red landscape hood seemed to be soft and pliable. The frame of the car now seemed to be responding to Hunter's will. For this split moment we were driving with two wheels.

By now, Anna was in the crash position, half-crouched in the back, her knuckles white from gripping the seat in front. Hunter kept pressing down, twisting the wheel from left to right, as though trying to force a knife blade between two looming surfaces. He was twisting the wheel like a leather whip and willing one of the red wings to bend up, over and fold onto its partner. The oncoming cars veered away — had to — but only the fact that there was a savage drop to a wet death strengthened their resolve to remain on what was left of a hard shoulder. The black truck lumbered on — the Red Shark morphed itself into a switchblade and slipped through the gap that wasn't there. It was not luck, nor good management, nor even divine intervention. The Red Shark was liquid. On a timeless thread, a gossamer wing, it took on a life of its own and animated itself to emerge like a smear of red cellulose into the open and on down the road.

'Crazy fucker!' mumbled Hunter darkly. 'Get the bastard's number, Ralph. We'll report him for dangerous driving!'

I looked back towards the trundling old truck with pen and notebook in hand, swaying a little, but in no danger. The driver was saying something himself, frowning and reaching for something. It looked like he was going to write something down too.

'I think he's going to take our number!' I said.

'Don't worry, Ralph! We have no plates on the back. Just a precaution. You can't be too careful these days. Fuck him, anyway! The bastard's spilt my drink! We don't need dangerous drivers in Aspen!'

Joe muttered disconsolately something about 'Bang goes another pair of decent pants.' And we made our way back to Owl Farm in silence.

READING *LONO*
OUT LOUD
2000

Plans for a Gonzo tattoo parlour . . . Limb-grafting . . .
Talk fat with cab drivers . . . The Curse of Lono *read*
on film . . . Labor Day upstream

Denver was becoming a home from home. Even the bellboys
knew us when we arrived at the Oxford Hotel, Stuart, who is
the warmest fan I ever met, and Dex and Rodolfo. They make
sure that we have everything we need and there is still a reassuring sense of tradition about the place.

We were there for the opening of another show of my
work at Bill Havu's new gallery, now the William Havu Gallery.
It was a large, modern building amongst the newly emerging
sporting condominiums, streets, office blocks, shops and all the
banal clutter of newness trying to be busy, and trying to find
its heart. The opening of the show was fizzing with an ener-
getic crowd and one guy asked me to autograph a huge tattoo
of one of the *Fear and Loathing* drawings already on his arm.
His name was Mitch and he had flown in from Minnesota
with his new partner, especially to have this done. He asked
me to sign it and went straight out to get it tattooed on proper.
He was a chef and a minister of the church and looked like
the actor Peter Boyle. His partner, whom he had met on the
Internet, owned a chain of toy shops. It's not the first time a
signing like that has happened. One guy had one that needed
signing on his leg. It looked like a nasty accident. Barstow bats
on the ass or the breasts are usually sent to me by email, merely
for approval, though I have been requested on occasions
through the years to sign around belly buttons, decorate spinal

tap enhancements, smear bare shoulders, delicately enhance feet and more intimate regions of the upper leg. Someone once wanted me to draw something on his face and sign it just below the nose but I had to decline that particular display of bad craziness in case somebody close to them noticed and sued me for defacing their loved one. I would not be happy if that was the first thing I saw every day for the rest of my life.

You have to be extremely careful, especially if you have had a few drinks and behave more involuntarily than usual. Gonzo is a strange kind of magic that appeals to the beast that lurks in the dark heart of most of us. A moment's frivolity could become a lifetime's regret. My best plan to date is to open a Gonzo tattoo parlour and make it an official business venture, with credentials, certificates of merit and risk indemnity. I would have to pass an examination, of course, to prove my intentions and if I took on apprentices they, too, would have to prove that they were embarking purely on an artistic endeavour. Collage Tattoo which is still in its infancy with metal inlays, bling crevice inserts and leather arm-welds would be attempted only under strict medical supervision. Total body transformation and limb-grafting is the Gonzo art of the future, as is decoration organ transplant which is, to date, the

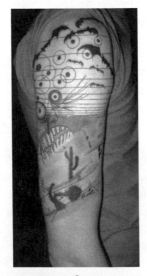

stuff of the mind. And nipple proliferation is unheard of until this moment.

There is nothing so pure as art and this was proved to me by the fact that when I was a very young man, artist's nude models would wander around stark bollock-naked, during ten-minute breaks, with a fag in one hand and a cup of tea in the other, perusing the student's faltering, flawed work their bodies inspired. One such model, whose name was Esther, often leaned over to inspect a particular detail of your drawing and would innocently prod you in the eye with a nipple. A man called Foster loved to present himself in a Charles Atlas pose in a half-kneeling posture. He would claim he could hold the position and then five minutes later would begin to grunt: 'Urgh! Arrgh!!'

'Are you allright Mr Foster?' the art teacher would ask.

'Yeaharrgh!' he would reply before collapsing in a heap on the floor. Seated poses were not his style and so he persisted throughout the lesson. It was a joy to have him there.

Most memorable of all was when I employed the great English eccentric Quentin Crisp, during my three years of teaching in the sixties. He was of slight physical stature and was possibly the most polite man I ever encountered. He could hold a still pose for half an hour without so much as a twitch. In break times he read studiously and never joined in any coffee breaks, preferring his own company as he sat, astride a stool, clad only in a limp, white jock strap. On one occasion I asked him if there was any pose of his own he would like to adopt. 'Certainly, Mr Steadman,' he would reply ever so politely. He was always the essence of polite decorum.

'How's this?' he said defiantly raising an oak chair, which must have weighed thirty pounds, above his head.

'Are you sure you can hold that, Mr Crisp?' I said.

'Yes!' he replied and he did, without so much as a tremble. Quentin Crisp was the sanest model it has been my privilege to employ.

I gave a lecture the next day in Bill's new gallery and I always like to talk through the images I have chosen to show using a slide-projector. I can hunker down and explain things

361

that cannot be explained in any other way. Perhaps that is when I am most truly myself. This night, a decent crowd listened attentively. One man was a studious fan called Gregory Daure or Gregory EGO, as he preferred to be called. We meet him every time we visit Aspen.

We were taxied to the gallery by a massively obese driver who obviously lived on burgers and coffee. Leftovers were strewn around the front of his cab and his weight had gradually distorted the back of his seat practically into our laps and we struggled to find space for our legs. He seemed welded to the upholstery like a Buddha.

Hunter's son, Juan, was at the gallery to greet us with his wife, Jennifer, and son, Will. Will interacted well with Hunter, who always insisted that his grandson should address him as Ace and not Grandpa, which would have sat on Hunter's shoulders like a massive goitre.

Age was not a friend to him. In fact, he treated it with great contempt as though it was something to be ashamed of, which of course it was, as his legions of friends discovered to their traumatic dismay. In the last few years it became ever more evident to me that my friend was suffering from the after-effects of his hip replacements, broken bones, invasive

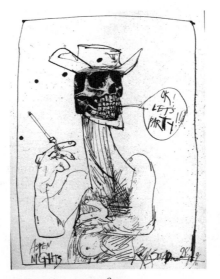

surgery, daily physiotherapy and prostate aggravation. His body was telling him things he didn't want to know.

On this visit, I had decided to get the entire text of *The Curse of Lono* read by all of Hunter's friends, since there was some doubt as to whether it would ever be published again. Hunter really appreciated having it read by his friends. He always liked to hear his words spoken. They came alive for him.

Even though it was Labor Day, 4 September, and he had hunkered down to watch the beginning of the football season, listening to the words took precedence. A brutal altercation had taken place the day before when Deborah had spent an enormous amount of time photocopying the unedited version of *Lono*. Hunter went ballistic and accused me, Deborah, Joe, and Anita, who had hardly yet had time to be a part of the domestic ménage, of 'feeding off him'. He said it often but the only reason to get this done was to prepare the book for republication sometime in the near future. My filming its reading was also a part of that.

Juan had arrived with Jennifer and Will, since we were all going up the hill to the home of a friend further up the valley, for the Labor Day celebration. Tragedy had struck one

year earlier when our hostess, Nicky, had rung Bob Braudis, Aspen's legendary Sheriff, in a terrible state. She had just found her husband, Oliver, dead in bed. Nevertheless, a year later, Nicky wanted to proceed with the celebration as her husband would have wanted it. In fact, I got much of *Lono* read to camera on that very day. Everyone was asked to wear a crazy hat and toast her late husband. In some ways it was surreal and wonderful.

Aspen is full of moments like that, as though life and death are interchangeable, charmed even, and we must go on. Maybe it is a pioneering thing. Most people who moved west in those days grasped at life as though each day might be their last. It was a moving experience because Nicky had insisted on the party being a life-affirming celebration of Oliver's life. It was a courageous thing to do. Something memorable. Hunter, too, was moved.

Back at Owl Farm I asked Juan if he would like to read the last few chapters of *Lono* for me and, naturally, he was a star performer, lending his reading something uniquely personal. For everything that Hunter was not, as a father, Juan made up for in forgiveness, understanding and a cherished respect for Hunter's obvious wayward frontier spirit. He never, in my company anyway, bitched or baulked or raged at a father who would have sent lesser sons into a protean brew of trouble-making resentment, a hell-king swine with a chip as big as the Rockies Great Divide. Juan had the birthright to slam back at everything and everyone and jam respectability and crony admiration back up the ass from whence it came.

Instead, he has shown a gentle regard for all who know him and love him. He displays a quiet charm, a generosity and an amiability that fly in the face of what may be expected of a noble demon's issue. He appears to have none of that arrogance and treats me as Uncle Ralph. I am honoured.

Anita, whom I had just met, has grown in our affections. The collective pain and anguish have given us a lot in common. For nearly thirty-five years I have endured, after unwittingly agreeing to meet him on his home turf, one of the most wanton, rebellious, dangerous and perfect creative collaborators I could have teamed up with, and a God-awful lot more he should have answered for. Instead, the cunning bastard checked out before he had to, leaving behind a battlefield of unexploded land mines, unused ammunition, guns, powders, salves, several bottles of the cheapest whiskies a self-proclaimed connoisseur would ever want to be seen dead with, uppers, downers, loofahs, quaaludes, a treasure trove of hilarious prose . . . but he left it to others to clear up the glorious mess.

THE LAST TRIP TO
WOODY CREEK
2004

*Ansaphones at dawn . . . Evil wears many masks . . .
Voting for Bush . . . Good people drink good beer
. . . Elvis leaves the building . . . Smoking back in
fashion . . . Looking for bears . . . Physio ain't
sport . . . Signing off*

In 2004, some time around early September, I received an answering machine message from Hunter that went on for five minutes. He rambled through several demands for things he needed from me for *Rolling Stone*. They were needed for the issue before the Bush–Kerry election – the election that broke his spirit, and the spirit of many others. I recorded his message and print it here as the last desperate call to arms from my friend who sounded, at the time, as if he was up against the wall.

Specifically, he wanted me to portray an image of absolute evil. 'Absolute' is an odd word and cannot be quantified, but when I have finished committing these words to print, I will try to do just that, in Hunter's honour.

Ah Ralph, you filthy little animal, filthy little beast, I have a job for you, a proud and noble job. I know, don't say it, this is just a job, another job. I am now writing on the fax, a rueful message, never mind, I am now writing, sending a page. Fax on the woeful message required, I'm a little high, Ralph, writing many weird pages on this article for Jann for this coming Friday, as my deadline, it's scheduled for the issue of October 1 2004 to the best of my knowledge. It's about voting and it's about elections − it's about vote or die − Ralph, this is about kicking ass, Ralph, and who else but you − who would I turn to when we want to kick ass − so, we will need some art on the US election, as it looks now. The real grit question for you, is this, for some reason, Ralph . . . er . . . what is the physical nature of evil? My real question for you, that I've written down here. Things that I had on my mind earlier. Yeh, then it occurred to me. For some reason this is what I wrote on my note book. Yeh, [a laugh] then a political drawing might stand out − with this art business − Somewhere in this context − your drawing of me in the jeep − We won't worry about the title or the caption yet [laugh] in that one − I am thinking of rewriting that *Time* magazine master-piece − it has become quite famous in the underworld of poof-poof journalism in politics − you know, the taco stand − yeh, all the time they're asking about you − I can probably get you a fellowship out here − imme-diately, come to think of it − now I understand you are coming here − so that's a different thing − of course you're coming here − when? Not soon enough to meet this deadline we have now in a week − for your take on my view on the eve of election − our view, Ralph, our view − yeh, fuck you, Ralph, oh and in the piece also which I sent you just − one lead, well I've sent you several leads, Ralph, call me − I look forward to the orgy − and believe me we're going to kick ass or get our asses kicked − before you get out of this country − if we do this art − er, we're just going to use the *Rolling Stone* conduit through October − to have an effect − and it's about the right time to do it − drop a bomb on the bastards − ah − so − do you have it? Ralph, you must have it, call me on − I'll sleep for a while after this message or maybe I'll swim − or maybe I'll go out and cruise the dark underbelly of Carbondale − who knows what I'll do? − but you're it. Thank you.

Tuesday, 7 September, 1.05 p.m.

Ah Ralph, one more thing concerning the art – ah, I'm going to send you, if I can find it – the absolutely classic political poster of Nixon – would you buy a used car from this man? – I am writing about it in this piece – I'm going to send it to you even on the fax or on the email – as soon as we find it – would you buy a used car from this man? – now showing Nixon, but Nixon was innocent compared to this man – alright, but also – I have a photograph – I'll send it to you, too – also – of a stripper, it was an orgy Ralph and I was present – and if you've seen this photograph, the Nixon art and the photograph of Cheney nuzzling a half-naked stripper – in Las Vegas – yeh, you'll see it Ralph, and thank you very much tonight.

Thursday, 9 September, 12.52 p.m.

Ah, Ralph, Ralph, let's see, I've got your drawings, I've done about 15 pages – on the piece but I want to know – hey, I don't know, Ralph, hey – something's wrong with my head – I'm seeing golf balls and little green men in yellow raincoats outside my door – and calling my name – anyway, are you talking to Jann, have you done anything? – the 'would you buy a used car from this man?' drawings are all good – emphasize – back on board – you'll be here soon enough – let's try to coordinate ourselves here – get this fucking thing done in a big way – make a serious splash of this – and I'm not sure the pig-fucking thing is the best way – I'm not sure myself – damn – I will be, so will you – give me a ring and how it looks and stands from your point of view. There's a certain amount of confusion between the deadline – the decision of what it is – dying – so, I left a message for . . . goddamn I wish I could work with children and animals – it's almost dawn here – And the first ray of the sun is coming up over the mountains, I may be out killing things. Whatever I can find. Allright . . .

It was a mischievous message with a hint of despair but I was up for it and I gave it a shot for I knew what he was referring to. The joke was indeed over and the fun had gone. It was a funny message, too, and tragic and George W. Bush will rot in hell. But I erred on the side of fun and created my image of a 'decent Republican'. Thick with irony, it was a stab at the worst and the most blatant, 'horriblest' creature that such a vision could invoke.

We would be in Aspen again in October when we could talk things over and I could work from there. First we would stay in Denver for a couple of days, visiting the Flying Dog Brewery, which had been using my artwork on its labels for some time. I had also been invited to attend the annual beer festival, which would also be Flying Dog's tenth anniversary. Volunteers would be encouraged to don plastic capes and submit to being spattered with red paint that I would hurl at them from the stage. I was to be driven out of a hangar backwards on a spiffy chopper motorbike by Eric Warner, one of the brewery directors. I had to trust him with my life and my drawing arm to be able to ride the goddamn thing. That part of our trip would be fun and we would stay at the Oxford Hotel again.

One of the great independent bookstores in America, the Tattered Cover, just around the corner from the hotel, had created a reading space for their readers' anti-Bush protestations and made comparisons between the purchase patterns of Democrats versus Republicans. Very kindly the general manager, Matt Miller, directed me to their site on the Internet. Though it might not have been completely accurate, here are some interesting comparisons, if the bookstore had guessed right. Democrats chose books like *Big Lies, Bushwhacked, Bush Women, Disarming Iraq, Downsize This! Dude, Where's My Country* — a Michael Moore must. They also chose *Stupid White Men, Lies and the Lying Liars Who Tell Them, The Best Democracy Money Can Buy, Thieves in High Places, Weapons of Mass Deception* and *Worse Than Watergate*. One detects a democratic flavour to all those titles.

There is a certain dismal self-regarding attitude in the

GREAT AMERICAN BEER FESTIVAL

Ralph STEADman 2004

ONE-EYED and SINGLE-MINDED.....!

titles presumably chosen by Republicans, such as *A National Party No More, Arrogance, Betrayal, Bush Country, Deliver Us from Evil, Hating America, Hillary's Scheme, Let Freedom Ring, Losing Bin Laden, Rumsfeld's War, Shut Up and Sing, The Faith Of George W. Bush, The Enemy Within, The French Betrayal of America, Things Worth Fighting For* and *The Right Man.* (Guess who that is!) They ooze Republican smugness. This may be completely misleading and a crazy kind of guess-work, but somehow I don't think so.

We spent time looking over the making of beer at the brewery. The best beer is made in micro-breweries like Flying Dog, which is one of the best but I would say that, wouldn't I? I am making a pitch for every little brewery and every little man in every endeavour of private enterprise. It is the way to go. These little guys are the new pioneers. I applaud them and loathe the corporate ruthlessness of massive take-over conglomerates that don't know what in hell they are producers of, except greedy figures to be slobbered over by fat-arsed shareholders who are interested only in the bottom line. If good people lose their jobs because these greed-heads want to see more profits, then damn their mindless mendacity and their total disregard for what should be of benefit to the entire community. I see a nation growing more distraught every time I come to the United States and each time the country is that little less united. Come out of your dump closets, you mean-assed Bushites who said before the last election: 'I won't be voting for Bush again.' Liars! Now look at the wretched state this world is in. Why in hell did you choose mean?

When I first came to America at the beginning of the seventies I was charmed by a certain naïve enthusiasm. I kept recordings from various radio stations to capture something of that naïveté. Only thirty-five years later, a disease has rotted the very heart of America that doesn't seem to want life and liberty anymore. America is ripe for lies and lethargy. The pure mountain air is going and gone. It is a huge burden and a sadness for us all.

Anyway, that's better. I need a bit of a rant from time to time. Just remember, as Hunter said: 'Good people drink good beer.'

We watched the installation of a new fermentation tank at Flying Dog and spent most of the day sampling the range and meeting the staff at the bar next door, where a Japanese film crew was doing its best to capture the living essence of a burgeoning industry. We played around a bit, hiding behind tanks and jumping out from odd places, saying: 'Boo!' It was accepted as common Western beer-drinking behaviour. We still had a party to go to where the founders of the brewery would meet up with us and celebrate the occasion. Eric Warner, George Stranahan, Richard MacIntyre and marketing manager Steve Charambulous, a young Englishman flexing his muscles American-style. As I say, I had to take Hunter's word for it that these were all good people.

The Denver Mayor, John Hickenlooper, was at the festival. He was after all also a founder member of the Wynkoop Brewery in lower downtown Denver, near Union Station. They were all there because it was the tenth anniversary of Flying Dog as well as the beer festival. So there were two events of some importance to good people who drink good beer.

The beer festival took place in a massive hangar. As

arranged, I was brought in on the bike and dismounted at the steps leading to the stage that was decked out in screens displaying the new beer, Wild Dog. I wore an apron and the helpers who were going to be splashed wore polythene cloaks. It was all very civilized and I was allowed to throw red paint around like a child.

Afterwards I found myself surrounded by six minders in white T-shirts with 'EVENT STAFF' printed on their backs. They accompanied me to a signing table and then stood around me, facing outwards like male caryatids with arms folded, to control the crowd, while I signed books and posters for the good people who formed an orderly queue. When that was done I was again surrounded by my minders in order to make my exit. I had never received such treatment before. Joe Petro, who had joined us for this trip, having taken all this in with some amusement, whispered into his cupped hands like a CIA official on an intercom system: 'Elvis is now leaving the building! Elvis is now leaving the building!' It was all a bit of a hoot, but my minders took their job very seriously and I thank them for it because, without them, I might never have got out.

Then we had Flying Dog's tenth anniversary party. We had a drink and then another one, the party wore on and suddenly I was in the middle of a room full of drunks. Thankfully, they were good, decent near-upstanding drunks. Everybody had a fine time and we provoked the gods by smoking heavily until dawn. It was the only decent thing to do. I look forward to the twentieth anniversary. Smoking will be back in fashion and respiratory side-effects will be a thing of the past. The smoke will no longer contain benzene, nitrosamines, formaldehyde or hydrogen cyanide and that may be a good thing or a bad thing. Who knows? Cars will be gone and rickshaws will be our preferred mode of transport. China will be in charge and America will be its main satellite country. Clean living and bottled water will be outlawed as part of an anti-plastic legislation and political correctness will be shoved into the long grass along with raccoon shit.

*

Back in Woody Creek, Anna, Joe Petro and Robert Chalmers and I stayed in George and Patti Stranahan's guest cabin. Our only warning was to watch out for bears. Okay. I remember the last time I had heard about the bears in Woody Creek. First, they were not a figment of the imagination and, secondly, they were now beginning to forage inside as well as outside houses as they had figured out how to open unlocked doors. One night about three years earlier, Hunter had heard one snuffling about outside his front door, had grabbed his shotgun and gone outside to investigate. Even on a moonless night in the snow, a big hulking shape is not an easy thing to miss. He had never wanted to hurt it but simply scare it away. He saw it and chose to aim at the ground around its feet and give it a jolt. Before he pulled the trigger, he warned Deborah, who occupied the cabin next to Owl Farm, to stay inside – 'There's a bear out here,' he mumbled – even when shouting, Hunter mumbled – and this alerted Deborah, who came out to see what all the commotion was. That was the moment when Hunter pulled the trigger and sent a Scotch mist of ricocheted lead flying all around the place. The bear bolted and disappeared but Deborah caught several of the pellets on her person and was hospitalized. For someone who always claimed to know about guns and how to treat them with respect, Hunter had a hell of a record of accidents to his credit. One could almost say that he was a benevolent mass murderer who wished no harm to anyone – but you never knew for sure in his world, so you had to stay clear of him and guns – or get involved and trust to luck. I did a drawing of the event showing a Baby Bear tugging at Father Bear asking: 'What did he do? What did he do??' Father Bear, who looks a little like Hunter replies: 'The little bastard shot me right in the ass for doing absolutely nothing – and that's about as good as it gets, son!!'

It was great to drive down to the Woody Creek Tavern and talk to living people again, even though they couldn't afford to live there any more.

Here is a list of all those familiar people who go to make up the real character of the most mythical bar on the planet. They are to be avoided at all costs or they will hit you for another drink. There's Michael Cleverly, artist and photographer; Gaylord Guenin, who has produced the best book about Aspen anywhere, called *The Early Years*; Cheryl Frymire, barmaid and friendly dispenser of the best service you could hope for; Cilla Hyams, an English rose who for years worked for Leon Uris; and Gerry and Chris Goldstein, who are dispensers of hospitality that begins and carries on non-stop. Then there is the massive, gargantuan giant amongst giants of a Sheriff, Bob Braudis, who believes that the law is to be used not as a weapon but a natural regulator of high spirits. Firm restraint is not a euphemism for clenching your buttocks so you can hardly breathe. Bob is the champion of the under-class and if he thought the occupants of a 4x4 Trouper Truck were drunk and stupid, rich and out of control, he was capable of turning the damn thing on its side, considering that to be a suitable punishment for crass dumbness and far more of a tutorial than a year in the slammer. Deedee, Bob's girlfriend, is like Mary Tyler Moore on speed. She insisted that I did tattoos on her boobs which, so I was told, took weeks to wash off and I was never arrested which proves what a reasonable man Sheriff Bob can be.

Never mind the immediate rich, high-flying neighbours like

Don Johnson and Melanie Griffiths, Henry and Jessica Catto, Sam Walton, Goldie Hawn and Kurt Russell, Michael Eisner, Robert McNamara, Michael Douglas and Zeta wotsit Jones, Jack Nicholson, George Hamilton, Donald and Ivana Trump, Jimmy Buffett, Barbra Streisand, Robert Wagner, Jill St John, Mohammed Hadid, Rupert Murdoch, for Christ's sake!, Prince Bandar, Sally Field, Martina Navratilova – is there no end to it and are they sending out a message that the rest just don't belong here? The air is too thin for the average mortal. The billionaires have been pushing out the millionaires for more than a decade.

One such billionaire was Floyd Watkins, who loved to hunt elk on the run from a helicopter. Hunter warned him he was fucking with the natural rhythm of the valley and then warned him off with a blast of shotgun pellets over his head. But people like that don't listen because they have impenetrable layers of bunkers, made of stashed money and battalions of paid lawyers who will fight to the death for their own personal cornucopia as though we still lived in the middle ages and the billionaires they work for are lawless barons.

Then there are the serfs – the Hispanic workers who, unless they are live-in domestics, have to live outside the town, on itinerant caravan lots with no toilets.

So much for that. It only embitters the spirit to dwell on others' good luck, even if they deserve it.

When we arrived at Owl Farm, Hunter was already up and the wretched sports channels were spewing out games nation-wide from every butch crevice that football can ooze from. This was Hunter's hideous obsession. The monster machine was never off. It glowed from the corner like a malevolent automaton. I tried to rationalize it.

Unless it spits blood, it ain't sport. Unless it gathers together the biggest bunch of blood-crazed bone-breakers inside a stadium fit for a Roman Empire to watch gladiators mangle each other into the dust, then it ain't sport.

The American psyche possesses an entrenched streak, forged in its national soul from the holy time when the Pilgrim Fathers left England in 1620 to worship in their own way. They came to a strange land and prayed with a passion that could not be sustained by a God that they could not see, or that they may have left behind.

That fervour has been transformed over three centuries into worship for another kind of god. A god of action – a superhero, or to be specific, teams of them. An American is born-again in a football stadium. The psyche's gods move fast and castigate the imagined enemies with a force and power sufficient to represent and evoke the force and power of the country itself.

American sport is an outward expression of the country's life-blood, the bursting of a dam, the outpourings of the reservoir of untapped energy. The driving force lies deep in America's psyche. Americans live with the certain knowledge that the source of their greatness has not yet been released. The intensity of their worship drives the gods in the arena beyond mortal goals and beyond mortal brute force, so much so that these gods are enshrined in superstructures to ward off evil spirits and the unthinkable possibility that their gods can be injured, albeit by another god from another place.

I can think of no other explanation for the hideous carnage that Americans demand from their sporting activities or the costumes that they fashion so ingeniously to adorn their gods. This peacock display of sporting *haute couture* defines their brashness and their ingenuity. Coupled with their talent for exploitation and big business, such a potent combination is not only irresistible to the American way, but to the rest of the world as well. Iraq has been one hell of a sporting event, but, unfortunately, there has been no end to it and even conservatives are beginning to quote Shakespeare and say: 'Enough! No more! 'tis not so sweet now as it was before . . .'

That is as much as I can say about sport. It is a generous view, and one that, I am sure, has never occurred to an American. It is a view that will not dim their enthusiasm nor dim the fact that sport drew some juicy pictures out of me.

*

'Oh, hi! You look tired. Make yourself at home.'

Hunter never said that — ever. He allowed people to fit into his world in the Owl Farm kitchen as bit-part players in a grander scheme of his own design; so you made up your own lines.

'You want whisky or something else — uh?' I had just walked into the kitchen and he was balefully indulging his greatest passion. But this time he was on his stomach, on a physiotherapy table, where he was receiving treatment to relieve the spinal pain resulting from his hip replacement as well as his painful spinal stenosis (where the bands of tissue that support the spine get thick and hard; the joints and bones enlarge and can bulge into bone spurs). This would have increased the already substantial pain that he suffered from the clinical arthritis I had noticed in his hands quite early on in our friendship. The daft bugger had also broken his left leg twice in the past two years, one of those accidents happening on a trip to Hawaii, or to be precise, in his case, a trip to the fridge. His strange body movements often put him off-balance and reaching for ice in the fridge had caused a broken ankle. He had been flown home by Sean Penn in a private jet and claimed that he had learned to walk twice in a year.

When we arrived, he had been picking at the desecrated remains of something that might have been breakfast but now looked like a dustpan of floor squalor ready for the trash can. He was looking a little paler than usual. The operations and leg-breaking habit were draining his usually vibrant constitution. After the physiotherapy he endured several times a week, he would stagger painfully back to his perch in front of his typewriter just as Anita got back from town with some oysters and various foodstuffs for the house and the bit-part players who consumed his day.

Other friends arrived — a local reporter called Troy and an older man called Tim Mooney who worked in real estate — and talk was of the imminent Bush–Kerry election. People would drop in from time to time, as though they were consulting the Oracle, which I suppose in a way they were. We were all handed Kerry buttons and info about the state of play. If you were a Bush supporter, then you just weren't there.

Hunter was still mucking with his food, picking something up and putting it down again and then looking at the TV game. He had installed an oxygen machine and we had a dose of that. God, it makes a difference at high altitudes. When Robert Chalmers turned up for a political interview, later to be called 'Day of Reckoning', for the *Independent on Sunday*, we were complete. During our stay, I went into the chemist store in Aspen and treated myself to my own oxygen canister and mask, which I shared with Robert, reckoning that he needed it more than me. The TV game was a good focus for the evening because, out of sheer perversity, I put forty dollars to win on the underdogs, Kansas City. I don't remember who they beat but Hunter wouldn't pay my winnings or even return the forty dollars up-front bet I had made because he was wrong and I realized he had been doing that to me for thirty-five years! I let it slip and forgot about it – but I never really did forget about it.

We watched the great debate between Kerry and Bush on video tape and stared in horror at this feeble-minded twit trying to take on John Kerry. I don't know how anyone could vote for such a man and many said they wouldn't, but as I have said before, many were lying. Later that night I read some pages out loud from Hunter's latest book, *Fire in the Nuts*, a hand-bound chapbook that I was publishing in a limited edition with Joe Petro and Walt Bartholomew.

We had another party the next night at the home of Gerry and Chris Goldstein, magical people we seem to have known all our lives. It was as though God had cut a lump out of the earth and set it down in Aspen and everybody just knew everybody else. It was never a problem to know people. It was as though Hunter, through the years, had cast a unifying spell over all who passed through Aspen, and through his life, on their way to somewhere else. Bob Braudis, the Sheriff, bear-hugged everyone who looked human and Deedee, his lady-love, gathered up the lost and made them secure in her nest until the next time. The feast was fit for kings and bums alike. All were welcome and Chris, Gerry's wife, was a great cook. The gods leave their spaces in the heavens and float down to earth to join the throng of Aspen inhabitants who believe that what happened in Aspen was normal and everyone else should have got in step. It was autumn – the first time we had been in Aspen when the leaves had turned, and the ambience was spectacular. The Aspen trees were the most intense yellow I had ever seen and in certain lights they glowed with an iridescent magic – nature's own theatre.

Getting Hunter to sign was, as ever, a problem. We had brought with us the limitation pages of *Fire in the Nuts*. We did not know it then, but it would be the last book of Hunter's published in his lifetime. However, he had signed too many times to be convinced that this was different to any other time, or that it was even in his interest to sign. Even I was a suspect – a sleazy no-hoper, to his mind, who was in his kitchen to suck from his leg-end – and I mean his leg-end – because everybody – even his friends – had sucked off him just one too many times. There was only one thing left to do – offer him money. Hunter loved money and if someone were prepared to part with the real boodle, in his hand, right then and there and make an Indian deal, he would go with that as an honourable contract. Instinctively, I knew that the only

way he would sign anything was to give him the money, so I made him out a dollar cheque. I remember him not quite believing it, even from me, but he agreed to sign the two hundred copies that were waiting for him at our cabin. I had already signed, and, so, I was giving him exactly what we would make if we sold them all. It was an offer he couldn't refuse and he always recognized a decent deal.

But can you imagine, at this moment, dear reader, that I am confident that all I am doing is coming into the home stretch, laying things to rest and sighing with relief that the bastard has laid himself down and gone to sleep? I am not writing about a serious writer as if he were some kind of ordinary person. No! He was, and can now be and unto eternity, a rabid, downright wretched, cheating, low-down sonofabitch, but that does not mean that he was wrong in his attitude to people or the fact that he wrote like an angel. We can no longer ask him to change a word, a phrase, a way of expressing a sentence. What he wrote is enshrined in death's immutable lexicon of useful things to be said at the right time, in the right place, and credit to the bastard for saying it first.

I waited for him later that night, bushy-tailed, willing him to turn up at our cabin like he had said he would and of course he did, but not on our time clock. He chose to turn up late with a raft of excuses, having had to travel only half-a-mile from where we were staying. He was lost, attacked by rabid livestock, driven silly with rage because he had a flat tyre again – he had been given a bad time by what he had heard on CNN and what we all knew to be his unnecessary postulation that he had fucked up and he was sorry that he could only make it at four o'clock in the morning. His bullshit was a wonderful aurora borealis of trepidation, failure, unnecessary hesitation and, something that no one but me knows, because he had previously confessed to me in one of those moments when all defences are down, that to do what was expected of him, officially, professionally and at a precise moment sent him into paroxysms of fear such as he was only able to express in point blank denials of ever being involved in something in the first place. It was his perfect and most indelible foible. I forgave him everything, admired his ability to play any system that could be manipulated. I deeply appreciated the pleasant charm that allowed him to saw through the bone of all the 'give me' hands that he cut off at the interosseous membrane or ligament of the forearm – the ulna – take what it was offering and walk off into the sunset, slapping both hands together and shrieking: 'Hot Damn!'

As I say, he never wanted to sign anything. 'People are always asking,' he said. 'I'll think about it. I'll come by later,' he added. We left it at that and returned to our cabin along the Woody Creek Road. By 3.30 there was no sign of Hunter. We decided to go to bed. But I knew Hunter would show up, so I left the pages to be signed on the kitchen table, with a pen. There was a commotion and horn-honking at about 4 a.m. I knew I was right. Later Hunter explained: 'There was a bear in the road.' I turned over and went to sleep.

I rose at 7.30 a.m. and walked sleepily into the kitchen. There was a polo bag on the table. Inside it was a half-drunk bottle of Chivas Regal, four boxes of cigars, a Gonzo thong, a gold krugerand and a magnum of a precious red wine, a fine and dignified cabernet sauvignon – nothing cheap! – which we shared with Hunter's lawyer, the Sheriff and friends that very next night. There were a couple of scribbled notes in Hunter's distinctive handwriting nearby. On one note he had written: 'Dear Ralph – Sorry I got lost in the night – I got a flat tire. Please help me to evaluate this profoundly rare wine. Love H.' On a second sheet he had written a list. 'Ralph – lettered sheets, numbered sheets. What else do you need? Ah yes – books signed, etc. – thank you. Hunter S Thompson.'

Bless him! he was going to do it!

He had clipped a smaller yellow piece of paper to the others on which he had scrawled 'You're welcome – the Fruit Fairy . . .' because he had stolen our cantaloupe melon.

So the pages had been gathered up and spirited away into the night like a guilty secret. My ole buddy would deliver, perversely, but he always delivered. A far more interesting signing than your average run-of-the-mill. Thought you would like to know that; it's a double first for a limited edition.

*

Walking back that night with Robert and Joe after a long hard day spent cross-examining Hunter for Robert's piece and assuring Joe that he would indeed get his signed limitation pages, we looked up at the moon. Joe looked off into the middle distance, which incidentally is nowhere in human terms, and said laconically, for it certainly wasn't enthusiastically: 'Y'know, I wouldn't mind betting that that is the last time we will ever see him.'

'Bullshit!' I replied. 'He hasn't even finished *Polo Is My Life*, and that is a must. I have the drawings to prove it!'

How wrong I was. Joe rang me at three in the morning on Monday, 20 February 2005, and warned . . . 'Take your phone off the hook. Hunter just shot himself.'

'About bloody time!'

Joe was right. The phone didn't stop ringing even when I had taken it off the hook.

His wife, Anita, is the sad, distraught torch-bearer for everything Hunter has ever done, ever engineered, ever manipulated, ever loved, ever given his attention to and ever fought for against injustice, calumny, greed and sloppiness. He was and is the enemy of stupidity, of brutality against the weak and silly. He stands for the antidote to the New Dumb.

Hunter S. Thompson was just another tax evader who got lucky.

MEMO TO THE
SPORTS DESK
Paris, 2006

Dear Hunter,

It is over a year since you set the scene
for your strangely planned suicide. The
messages that I had received from you up to
the Bush-Kerry election of 2004 were infected
by a dismal undertone. 'If Bush wins,' you
said, 'the planet is doomed. The Halliburton
Corporation will rule with an army of obedient
geeks who will transform America into a land
fit for religious freaks, informers, fascist
legislators, greedheads, an obscene defence
budget against the 'terrorists' and the desire
for pre-emptive action against Iran and the
ambition to remain the most dominant power on
earth by whatever means.' The last message
you left me asked me if there were a way I
could draw the personification of absolute
evil. It was a tall order and I tried several
images but your futile plea had eluded me.
Maybe I am looking in the wrong place!

It was a conundrum to be solved after you
were gone, though I did not realize it when
we were in Woody Creek in October 2004. We
were in George Stranahan's ranch house and on
the kitchen table were my inks, paints and
pens ready to do justice to your request. I
did not suspect that this would be your last.
Why shouldn't I think that this was one of

those times when I could deliver like before,
do the proper thing? Kick ass or get kicked
in the ass? Well, we got our asses kicked.
The world got its ass kicked. Although I tried
to give my best on this occasion it wasn't
enough. Cartoon satire was not a cure-all
answer to anything, particularly not regarding
the election of the President of the United
States. The President embodies all that is
wrong with our planet, our environment and
our predicament and your beloved nation was
duped into voting for him by corporate deceit.
Bush is the first President to do this (save
for our old friend Richard Nixon whom you had
begun to consider as a barrel of laughs compared
to this one).

George W. Bush is a fraud who beat Al Gore
by cheating and won a second term by presenting
his credentials in a violent lie as the enemy
of terrorists and the saviour of democracy
throughout the world. He is scum and always
will be scum. He will live on as the lowest
form of intelligence at an executive level
this world has had to suffer.

Hunter, that is why you committed suicide.
Your America had gone. It was seriously the
death of fun. THE JOKE WAS OVER. When you
pulled the trigger on your Magnum .44, having
invited the one man in the world you could
trust to be there for you, your son Juan, you
blasted away one of the unique brains of our
era. It was a good shot which not only took
your brain but pierced your cooker hood as
well. A ventilation shaft to eternity. God!
We should feel good but we don't. Most who
knew you, Hunter, are bereft and rue the day,
that day my working partner of thirty-five
years blew himself away.

My first reaction was to say, 'About bloody
time! He's been threatening to do it for years.'

'Take the phone off the hook,' Joe Petro
had said. But next morning I put it back on
its hook and luckily my first call was from
The Independent, who asked me to write a lead
article on my association with you. Headline
news! Not only headline news but headline
headline news.

Over the next few months I received many
requests from broadcasting companies. They were
like wasps sucking the sugar and the protein
to feed their stories. Each had plenty of
reasons why I should tell them everything.
Yet with each call I felt less urgency to
provide them with what they desired. I became
indolent, taciturn, even uninterested. Of
course it was a story to pursue and be fired
up by, but to me it had become a tragic replay
of something I had hoped would never happen.

I have re-enacted the moment when you
convinced yourself that this was it. Did you
think about never seeing familiar things again
— never being cruel, never laughing, never
again having a single thought, never knowing
another spring, never lying, never reading a
truth, never having an opinion, never railing
against the New Dumb, never figuring out that
there was still fun to be had, good pain to
be endured, love to be embraced, never again
to gather in wood for the winter months, kick
the snow, feel the chill, enjoy the cold
moment, gather in the elements that express
the signs of life, be Gonzo, enjoy the response
of surprise that gives Gonzo its expressive
truth?

Never to know ever again that life is a
drug. Life is what gives life a reason to try

out of sheer curiosity. The smash of a bullet
through that sentient organ which contained
everything you had experienced, known, remembered, tortured yourself with, gained to your
advantage, gathered, stored, evaluated,
resisted or returned.

Before the bullet a terrified moment of
resistance, a fleeting cry for help — why me?
Now? No! No! Now! — knowing that your death
was a squeeze away, a chasm between here and
oblivion.

Maybe you reached eternity but the gods
were out to lunch. The limitlessness of time
lends lassitude to the gods and they don't
give a shit. You defied them. You laughed at
them and scorned their efforts to bring you
to your knees when you were young.

Given their normal anger they had been kind.
You provoked all and everything as part of a
game that was never the same to them. Some
subterranean threat, a massive plunge of
unearthly shock to the vital depths of your
being, lay waste the simple choice between
life and death. To lesser mortals it was only
a twinge but to you it was a death blow. You
were never reasonable and you knew, in spite
of your wayward spirit, you were not invin-
cible. The moments of doubt multiplied and you
pitched your bravado against the phantom enemy.

Those who grew to be a threat to your
America continued to burgeon. You faced them
and made your choice. The brutal contempt of
the majority convinced you to let it be, throw
in the towel and give up the ghost. But, for
you, it was victory or game over. I just
wanted you to hang around and nail the beauty
of that image of 'absolute evil'.

Your spirit has prevailed and dominated the

conscious lives of many. You might even have
lived to see a day when the most conservative
spirit would have joined you to eradicate the
congenitally bad and embrace the wickedly good
that made life such fun. Characteristically
you chose to pre-empt the impossible in favour
of the possible. You carved a place for your
massive dream and left no space for those who
would fight on.

But you leave us with a blueprint, ole
sport. Take it up with the gods. Send word.

Ralph X

INDEX

Acosta, Oscar 66–7, 68, 69, 70,
    74–5, 155, 169, 177
Agnew, Spiro 91
Ali, Muhammad: 'Rumble in the
    Jungle' 115, 116, 117, 121, 124,
    126–30
Allen, Paul 160, 314
Allende, Salvador, President of
    Chile 101
America's Cup, the (1970) 45–58,
    62–3
Amps, Kym 284
Andoe, Joe 316
Aspen, Colorado 339, 355, 374–5,
    378–9
    airport 138–9
    British Design Week (1986)
        337
    Carriage House 318
    Steadman's exhibition (1996)
        337, 338, 341
    Thompson's Sheriff campaign
        39, 40–41, 58, 82, 83–4
    see also Woody Creek
Auman, Lisl 68

Bacon, Francis 327
Bailey, Andrew 124
Ballantine, Ian 245, 247, 250, 254,
    258, 259, 260–61, 262
    Steadman's letter to 263–7
Bandar, Prince 375
Bantam Books 191, 245, 255, 262,
    267–8
Bardstown, Kentucky 310

Jim Beam's Distillery 310
Oscar Getz Museum of Whiskey
    History 310
Bartholomew, Walt 378
Basalt, Colorado 355, 356
Bass, Saul 337
Belushi, John 112, 256
Beneduce, Ann 60–62
Boston Globe, The 69
Boulder, Colorado: Clarion Hotel
    321–2, 324
Boyle, Peter 177
Bradley, Ed 318
Braudis, Bob, Sheriff of Aspen
    323, 324, 364, 374, 378
Brown, Bundini 124–5
Brown, James Graham 309
Brown Hotel, Louisville 309–10
Brown's Hotel, Mayfair 134, 135,
    298
Buchanan, Pat 342
Buffett, Jimmy 375
Burroughs, William 316, 325–6,
    327–30, 341
    The Proud Highway 325
Bush, President George W. 89,
    369, 377, 378, 383, 384
Butterfield, Alex 97, 163

Captain Steve 199–200, 207–8,
    209, 210–11, 212, 213, 215,
    217, 218, 219
Cardoso, Bill 69, 119–20
Carter, President Jimmy 100, 171,
    227

Cassady, Carolyn 326
Cassady, Neil 326
Catto, Henry and Jessica 375
Causley, Charles 327
Chalmers, Robert 297, 298, 299,
    300–301, 302–3, 373, 378,
    382
Chapin, Dwight 163
Charambulous, Steve 371
Chelsea Arts Club, London 136
Chicago Riots 82
Clancy, John 94–5
Cleverly, Michael 374
Clinton, President Bill 309
Colson, Chuck 101–2, 106, 163
Conrad, Joseph 270
Corso, Gregory 326
Crisp, Quentin 361
Crouse, Tim 89–90

Daily Telegraph 204–5
Daly, John 123
Daure, Gregory (Gregory EGO)
    362
Dean, John 97, 161–3, 173
    Blind Ambition 162
Dean, Mo 173, 174
Delise, Don 318
Democratic Convention (Miami,
    1972) 84–5, 87, 88–9
Dempsey, Michael 41, 256
Denver, Colorado
    beer festival 371–2
    Flying Dog Brewery 369, 370,
        371, 372
    Havu Gallery 359, 361–2
    One on One gallery 326, 336
    Oxford Hotel 326, 337, 359,
        369
    Tattered Cover bookstore
        369–70
    Wynkoop Brewery 371
Denver Post 9, 39
Dibb, Michael 202
Didion, Joan 95

Diller, Barry 305, 306, 307, 308
Dix, Otto 85
Don't Tell Leonardo (film) 202
Douglas, Michael 375
Douglas-Hume, Charlie 37–8
Duffy, Carol Ann 327
Duncan-Sandys, Laura 286
Durrell, Lawrence 64, 327

Edinburgh Festival (1987) 272–3
Edward (Thompson's mynah bird)
    110–11
Ehrlichman, John 97, 104, 107, 163
Eisner, Michael 375
Ellsberg, Daniel 6, 107–8
Ervin, Sam 99, 103, 104–5, 107
Ewing, Wayne 356
    Breakfast With Hunter 356
Exeter Arts Festival (1989) 283,
    285

Fainlight, Harry 327
Fainlight, Ruth 327
Falklands War 252, 254, 256–7
Faulkner, William 69, 270
Fear and Loathing in Las Vegas
    see under Thompson, Hunter S.
Fear and Loathing in Las Vegas
    (stage version) 237–44
*Fear and Loathing on the Road to
    Hollywood* (*Omnibus* film)
    137–43, 144–61, 163–6, 168
Ferlinghetti, Lawrence 326
Field, Sally 375
Financial Times 185, 186
Finch, Nigel 137–41, 142, 145–6,
    147–52, 154–5, 156, 160, 163,
    166
Florian, David 316
Foreman, George 115, 116, 123,
    126–30
    Steadman's interview 121–2
Foster, Stephen 310
Fox and Anchor Pub 299
Frazier, Joe 127

Freud, Sigmund 185, 188–9, 327
  *Jokes and Their Relation to the
    Unconscious* 188
  Totem and Taboo 222–3
Frost, David 123, 135
Frying Pan River, Colorado
  353–4
Frymire, Cheryl 374
Fuller, Deborah 317–18, 320, 321,
  333, 341, 342, 346, 355, 363,
  373

Garfunkel, Art 94, 108
Garfunkel, Linda 108–9
Gate Theatre, London 240–44
General Election (1970) 36–7
Ginsberg, Allen 326, 329
Glazer, Milton 337
Gleason, Ralph 67
Glenwood Springs, Colorado 82
Goddard, Donald 9, 45
Goddard, Natalie 9
Goldstein, Chris 318, 374, 378
Goldstein, Gerry 318, 320, 331,
  332, 341, 342, 343, 374, 378
Goldwater, Barry 91
Gollins, David 186
Gonzo 69, 360
  25th anniversary celebrations
    330–35
Grauerholz, James 328
Greath, Bill 306
Gregson-Williams, Richard 283,
  284, 285
Gretel II (yacht) 48, 51, 55
Grey, Patrick 107
Griffiths, Melanie 375
Grosz, George 24, 85
Guardian 278, 280
Guenin, Gaylord: *The Early Years*
  374
Gysin, Brion 326

Hadid, Mohammed 375
Haig, Alexander 205, 227

Haldeman, Harry R. ('Bob') 97,
  104, 106, 163
Hamilton, George 375
Harvey, Richard 283, 285
Havu, William 326, 336, 359, 361
Hawaii 220–21
  City of Refuge 221–3
  see also Honolulu Marathon;
    Kona
Hawn, Goldie 375
Heaney, Seamus 327
Hearst, Patty 95
Heath, Edward 22, 35, 36, 37
Hemingway, Ernest 170, 210,
  270
Henri, Adrian 64, 327
Hickenlooper, John, Mayor of
  Denver 336, 371
Hinckle, Warren, III 6, 34, 42, 45,
  79
Hinckley, John 204
Hockney, David 337
Holm, Ian 284
Honolulu Marathon (1980) 184,
  191, 195–6, 206–7, 227
Hughes, Ted 64, 327
Hyams, Cilla 318, 374

Independence Pass, Colorado 82,
  354–5
Independent 45, 281, 296, 385
Independent on Sunday 378
Indica Bookshop, London 326
Intrepid (yacht) 48, 51, 55

Johnson, Don 375
Jones (Thompson's cat) 111–12
Jones, Catherine Zeta 375

Kalmbach, Herbert 97, 163
Kansas City: Republican
  Convention (1976) 97, 162
Kelner, Simon 45, 272, 296, 300,
  301, 302, 303, 304
Kennedy, Edward 91

Kentucky Derby (1970) 21–8, 69,
    344; (1995) 331–2
Kerouac, Jack 325, 326
Kerr, Gordon 311–12
Kerry, John 377, 378
Khaled, Laila 59
King, Don 126
Kingsley, Ben 284
Kona, Hawaii 196–7, 204, 206–7
    Gold Jackpot Tournament
        (1981) 207–20; (1982) 248
    Huggo's Bar 197, 208, 211, 230
Kramer, Lyn: *Freud* 327

Lagos, Nigeria 130–32
Laila (Thompson's girlfriend) 112,
    186, 187, 190, 191, 195, 199,
    200, 202, 228, 230, 248, 294,
    320, 321, 322
Laird, Melvin 104
Las Vegas 68, 70, 145, 146–7
Latchmere Theatre, London
    237–40
Leadville, Colorado 82, 350–52
Lee, Laurie 327
Leibovitz, Annie 107
Lennon, John 207
Leonardo da Vinci 160, 185, 201–2,
    258
    The Last Supper 202, 258
Lexington, Tennessee 312–14
Linden, Eddie 327
Linson, Art 161, 168–72, 177
Liverpool: World War II bombing
    raids 86
Lobner, Hans 189
Lock, Martin 186
Logue, Christopher 64, 327
London Marathon 204–5
Loretto, Tennessee: Maker's Mark
    Distillery 310–11
Los Angeles 147, 158, 159–61
    Universal City 168
Los Angeles Times 67
Louisville, Kentucky 309

Luxor, Egypt 284
Lynchburg, Tennessee: Miss Mary
    Bobo's Boarding House 311

Macbeth, George 327
MacGibbon and Kee 41
McGovern, Eleanor 99
McGovern, George 87, 98–100, 101
MacGruder, William 163
McIlvanney, Hugh 130
MacIntyre, Richard 371
McNamara, Robert 375
Mailer, Norman 130
Malanga, Joe 326–7
Mee, Margaret: *In Search of the
    Flowers of Amazon Forests*
    283–4
Metnik, Sam 191, 252
Metropole Hotel, London 298,
    299–301, 302
Miami, Florida 87–8
    Democratic and Republican
        Conventions (1972) 84–5,
        87–90, 91, 100
Miles, Barry 326
Miller, Henry 327
Miller, Matt 369
Minihan, John 327
Mitchell, Adrian 327
Mitchell, John 163
Mobutu, Joseph, President of Zaire
    123, 126, 127–8, 130
Mooney, Tim 377
Murdoch, Rupert 375
Murray, Bill 163, 171, 176, 179, 249

Nabulsi, Laila *see* Laila
Nashville, Tennessee 311–12
Navratilova, Martina 375
Nesbit, Lynn 40
New York Times 6, 9, 107
New Yorker 305, 307
Newport, Rhode Island 45, 46, 47,
    50–52, 82
Nicholson, Jack 375

Nicole (Thompson's assistant) 297,
    298, 299, 300
Nixon, President Richard 6, 100,
    157, 171, 177–8
    and John Dean 97, 161, 162, 163
    death 308
    Thompson's opinion of 70, 91,
        308, 342, 368, 384
    and Watergate 97, 100, 101, 104,
        106, 107

Observer 272, 296–7, 302, 303, 304
October Gallery, London 327
Oddbins 272, 309, 311–12
Oliphant, Pat 9, 39
Omnibus (BBC) 137, 138
O'Rourke, P. J. 302
Orwell, George: *Animal Farm* 269
Owl Farm *see* Woody Creek

Paepke, Mrs Walter 83
Paisley, Ian 37
Parkinson, Norman 337
Patten, Brian 64, 327
Pendennis, the (Louisville club)
    28–30
Penn, Sean 377
Perles, Alfred 327
Perry, Paul 292
    *The Strange and Terrible Saga
    of Hunter S. Thompson*
    292–5, 295
Petro, Joe, III 312
    in 'Burroughs Gang' 329, 341
    at Denver beer festival 372
    on fishing trip 353–4
    and Hunter's death 382, 835
    publishes *Fire in the Nuts* wit
        with Steadman 378
    silkscreen printing 313–14
    at the Stranahans' 373
    on trip to Basalt 356, 357, 358
    at 25th anniversary celebrations
        of birth of Gonzo 330, 331,
        332, 333, 335

    visits Hunter's mother 344, 347
Plath, Sylvia 64, 327
Plimpton, George 130
Poe, Edgar Allan: *The Imp of the
    Perverse* 205
Polky (Petro's girlfriend) 312, 313,
    330, 331, 332, 333–4
Porter, Peter 327
Powell, Enoch 36–7
Private Eye 38
Punch 279, 281

Radio Times 38
Ramparts (magazine) 107
Random House 76, 80
Rattiner, Dan 7–8
Rattiner, Pam 7–8
Ray, Man 327
Reagan, President Ronald 203–4,
    205
Reed Brothers (undertakers) 164
Rees-Mogg, William 37
Reeves, Mike 164–6
Republican Convention (Miami,
    1972) 84–5, 89, 100
Rhodes, Zandra 337
Richards, Keith 160
Rifle, Colorado 338, 339
Rinzler, Alan 225
    Thompson's letter to 191–4
Roaring Fork River, Colorado
    113–14
Robinson, Bruce 281
Rockies, the 82, 338–9
Rocky Mountain Airways 138, 321,
    336, 338
Rogers, Deborah 237
Rolling Stone magazine 30, 67–8,
    69–70, 71, 73, 74, 75–6, 78,
    79, 81, 90, 91, 93–4, 95, 97,
    99, 107, 115, 124, 136, 167,
    168, 179, 281, 316, 366, 367
Running magazine 195–6, 206
Russell, Kurt 375

St John, Jill 375
Salazar, Ruben 67–8, 69, 70
San Francisco 92–5, 106–9
  City Lights Bookshop 326
Scaggs, Boz 95
Scanlan's magazine 6, 8–9, 10, 22,
  33, 38, 42, 45, 62, 69, 79
Scarfe, Gerald 306
Seattle, Washington 160, 314
Segretti, Donald 97, 163
Silberman, Jim 76, 147, 150
silkscreen printing 313–14
Sillitoe, Alan 327
Simon, Paul 94
Sly and the Family Stone 94
Smith, Patti 326
Southern, Terry 326
Spender, Stephen 64, 327
Sports Illustrated 68–9, 70
Stans, Maurice H. 163
Steadman, Anna 38, 63, 65, 203,
  226, 295, 296
  in America with Steadman 30,
    92–3, 96, 99, 106–9, 168,
    312, 313, 333, 344, 347
  on fishing trips 114, 353, 354
  in Hawaii 195, 199
  and Hunter 187, 280, 315, 321,
    356, 357, 358
  and Juan's wedding 321–2, 324
  in Luxor 284
  at Owl Farm 156, 318, 320,
    332, 333
  in Vienna 189
Steadman, Henry 38, 93
Steadman, Ralph
  books:
    Alice in Wonderland 40
    Alice through the Looking Glass
      40, 41, 45, 63, 64, 72, 79
    The Big I Am 269
    Fear and Loathing in Las Vegas
      see under Thompson, Hunter
    Gonzo the Art 316–17
    I, Leonardo 201–2, 258, 268, 314

Orwell's Animal Farm 269
Polo is My Life see under
  Thompson, Hunter
Sigmund Freud 185, 188–90
Still Life with Raspberry 9
cartoons and drawings:
  The Babel Tower and the
    Attempted Assassination of
    Ronald Reagan 202–3
  Bob Dylan 178, 179
  Early Morning Scene 36
  Gonzo Guilt 172
  The Lie Detector 106
  Nancy Reagan 203
  Nixon Crucified in a Stained
    Glass Window 107
  Spirit of Gonzo 168
  Still Life with Bottle 311
  'Today's Pig is Tomorrow's
    Bacon' 278
  The Unhappy Clown 36
  The Wasteland 36
exhibitions: 326, 336, 337, 338,
  341, 359, 361–2
poetry and songs:
  'Friends' 281–3
  Plague and the Moonflower
    283–5
  'Those Weird and Twisted
    Nights' 173–6, 178
prints:
  Bats over Barstow 313
  Burroughs 330
  Lizard Lounge 313–14, 347,
    348
  Shot Sheriff 329, 330, 332
  Vintage Dr Gonzo 329, 330,
    332, 333–4

Steadman, Sadie 168, 187, 195,
  197, 198–9, 246, 251, 261
Steadman, Theo 38, 93, 98, 252,
  254, 255
Stein, Abner 72, 76, 255
Stein, Lou 237, 239–40, 242, 243

Stephens, Olin 51
Stirling, James 337
Stirmer, Dougall 107
Stone, Bernard: bookshop 63–4,
    65, 136, 187, 252, 326–7
Stranahan, George 355, 371, 373,
    383
Stranahan, Patti 355, 373
Streisand, Barbra 375
Studt, Richard 284
Suares, J. C. 8–9, 34, 45–6
Sunday Times 38

Tabor, Horace: Leadville Opera
    House 352
tattooing 359–61
Tax Loss (group) 178
Thatcher, Margaret 233, 234
Thomas, Justice Clarence 305
Thompson, Anita 355, 363, 364, 3
    77, 382
Thompson, Davison 14–15, 323, 3
    46
Thompson, Hunter Stockton
    and Acosta 66–7, 68, 70, 74–5,
        169
    alcohol consumption 13, 129,
        170, 298, 299–300
    and America's Cup (1970)
        45–8, 49–58, 62–3
    appearance 12–13, 297
    and Burroughs 325–6, 329
    cars and driving 14, 101, 119,
        139, 153, 156–7, 159, 315– 16,
        355–8
    and Chalmers 297–303
    childhood 2, 199, 353
    and children 198–9
    and computers 187–8
    'conceptual schizophrenia' 68, 70
    as crusader against injustice
        66–8, 81, 335
    drugs 46, 51, 124, 135–6, 158–9,
        179–81, 297, 299, 301
    eating habits 19, 271, 312, 378

    and Edinburgh Festival 272–3
    and the 'Fear' 245, 246
    and filming of Fear and
        Loathing on the Road to
        Hollywood 137, 139–43, 145,
        146–56, 160–61, 163–6
    and fireworks 199
    first meeting with Steadman
        11–12
    and Gonzo journalism 69
    and grandson 362
    and guns 17, 86–7, 139–42, 277,
        319–20, 332, 333, 373–4
    in Hawaii 184, 195–200,
        206–24, 227–8, 248–51
    and Juan see Thompson, Juan
    and Juan's wedding 320–24
    Kentucky Derby and after
        20–22, 24, 25–6, 28–31,
        32–3, 38–9, 41, 69
    kleptomania 168–9, 179
    last telephone messages 366–8
    lawsuit against Scanlan's 79
    Legal Defense Fund needed
        278–81
    letters and faxes to Steadman
        38–40, 41–4, 75–6, 78, 80,
        123–4, 226–9, 260–63,
        276–80, 285–94, 303–9
    Memorial Blast-Off xiii, 1–2,
        95, 164–6
    nasty habits 109–10
    national pride 143
    and New Dumb 157, 188, 382
    and New York Customs 132–4
    and Nixon 70, 91, 308, 342,
        368, 384
    and Observer articles 296–304
    pets and peacocks 110–12, 186,
        318, 332, 355
    physical ailments 17, 201,
        362–3, 377
    and politics 82–3, 98, 101–2
    relationship with Steadman
        115–16, 138, 144, 146, 169–70,

184, 186–7, 190, 245, 269–70,
271–2, 278, 279, 294–6, 335,
337, 341–2, 364–5
and *Rolling Stone* magazine
67–8, 69–71, 78
and shady people 197–8
Sheriff of Aspen campaign 39,
40–41, 58, 82, 83–4
signing books and prints 2,
330, 332–4, 379–80, 381
and Steadman's 'ecological' book
79–82
and Steadman's son 98, 230–36
and Steadman's whiskey trip
309–10
suicide xiii, 2, 382, 383–7
and television sports channels
375
trial for drunk-driving 342–3
and 25th anniversary of birth of
Gonzo 330, 331–3
unpunctuality 380–81
writes foreword to Steadman's
Gonzo the Art 316–17
in Zaire 118–21, 122, 123–6,
128–32
works:
Better Than Sex 289–91
The Brown Buffalo 155
'The Charge of the Weird
Brigade' 245
The Curse of Lono 2, 190–94,
195, 200–201, 204, 206, 225,
226, 228–30, 245–6, 247,
255, 258, 260–68, 301, 363,
364
Fear and Loathing in Las Vegas
66–74, 75–7, 93, 184, 231,
313–14
Fire in the Nuts 378, 379–80,
381
The Great Shark Hunt: Volume I
187, 188
Hell's Angels 188, 231
'I Told Him It was Wrong' 64–5

Polo is My Life 316, 335, 340,
382
The Rum Diary 68
Songs of the Doomed 147
Where the Buffalo Roam (film)
161, 163, 167–73, 176–8, 179
Thompson, Jack R. 346
Thompson, Jennifer (*née* Winkel)
316, 320, 322, 323, 362, 363
Thompson, Jim 346
Thompson, Juan
character 93, 323–4, 364
childhood 93, 94
and Hunter 112–13, 114, 145,
155, 197, 233, 261, 324, 364
and his mother 82, 247, 323–4
at Steadman's exhibition 362
363
wedding 315, 316, 320, 322–4
Thompson, Sandy 82, 96, 99, 113,
145, 151, 154–5, 247, 323, 324
Thompson, Virginia 344–8
Thompson, Will 362, 363
Thorpe, Jeremy 36
Time magazine 367
Time Out 237, 240
Times, The 35, 37, 45, 62, 63
Trombley, Stephen 281
Trump, Donald and Ivana 375
Tullahoma, Tennessee: George
Dickel's Distillery 311

Uris, Leon 374

Vienna: Freud Museum 189
Vietnam War 99, 107, 116
Veterans 87–8

Wagner, Eric 371
Wagner, Robert 375
Walton, Sam 375
Warner, Eric 369
Washington, D.C.: Hilton 96, 101
Watergate 96–7, 99, 100–7
Watkins, Floyd 375

Wenner, Janie 93, 94, 107, 108
Wenner, Jann 67, 70, 75, 76, 84,
    85, 89, 90, 92–3, 94, 107–8,
    168, 178, 276, 306, 366
Whitmer, Peter 292
Whitmire, Carrol 82
Wilbur, John 206, 208, 251
Wilde, Oscar 82, 352
Williams, John 283, 285
Willner, Hal 328
Wilson, Harold 35, 36
Wiseman, Octavia 76
Woody Creek 82, 316, 343, 355,
    373

Hunter's Memorial Blast-Off
    xiii, 1–2, 95, 164–6
Owl Farm 109–12, 134, 198,
    316–17, 335–6, 355,
    377
Tavern 41, 139, 316, 331, 335,
    374
Wyckoff, Barney 339, 341
    gallery 337, 338, 341

Zaire: Ali–Foreman fight
    115–30
Zion, Sidney 6, 34, 45, 79